Microarrays for an Integrative Genomics

Computational Molecular Biology

Sorin Istrail, Pavel Pevzner, and Michael Waterman, editors

Computational molecular biology is a new discipline, bringing together computational, statistical, experimental and technological methods, which is energizing and dramatically accelerating the discovery of new technologies and tools for molecular biology. The MIT Press Series on Computational Molecular Biology is intended to provide a unique and effective venue for the rapid publication of monographi, textbooks, edited collections, reference works, and lectures notes of the highest quality.

Computational Molecular Biology: An Algorithmic Approach
Pavel Pevzner, 2000

Computational Modeling of Genetic and Biochemical Networks
edited by James Bower and Hamid Bolouri, 2001

Current Topics in Computational Molecular Biology
Tao Jiang, Ying Xu, and Michael Q. Zhang, editors, 2002

Gene Regulation and Metabolism: Postgenomic Computation Approaches
Julio Collado-Vides, editor, 2002

Microarrays for an Integrative Genomics
Isaac S. Kohane, Alvin T. Kho, and Atul J. Butte, 2003

Microarrays for an Integrative Genomics

Isaac S. Kohane, Alvin T. Kho, and Atul J. Butte

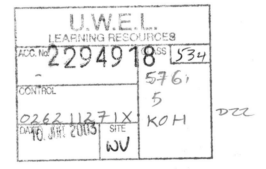
A Bradford Book
The MIT Press
Cambridge, Massachusetts
London, England

This book was set in in Computer Modern by the authors using the LATEX typesetting system and was printed and bound in the United States of America.

Library of Congress Cataloging-in-Publication Data

Kohane, Isaac S.
Microarrays for an integrative genomics / Isaac S. Kohane, Alvin Kho, Atul J. Butte
    p.   cm.—(Computational molecular biology)
    Includes bibliographical references.
    ISBN 0-262-11271-X (hc.: alk. paper)
    1. DNA microarrays. 2. Gene expression—Analysis—Automation. 3. Bioinformatics. I. Kho, Alvin. II. Butte, Atul J. III. Title. IV. Series.
QP624.5.D726 K686 2002
572.8′6–dc21

                                                            2002022663

Contents

Foreword

The impact of microarray measurements on biology and bioinformatics has been astounding. Starting from virtually no literature a few years ago, this field has come to dominate many conferences and journals. As an example, the Intelligent Systems for Molecular Biology conference, the annual meeting of the International Society for Computational Biology held in Copenhagen in 2002, had almost 50% of its papers in the areas addressed directly or indirectly within this book. Four years ago, there were none. Bioinformatics has always been driven by the availability of data—sequential, structural, and most recently functional. The availability of sequence data brought into biology a cadre of computer scientists with special skills in string processing. The availability of structural data brought in technical experts in visualization and computational geometry. This most recent development—the availability of relatively large data sets measuring the expression of genes within cells—has helped attract yet another group of scientists—the data miners, machine learners and statisticians.

In many ways, the impact of this data on biology and informatics can be summarized in figure 1.4 of this book—the world has not quite been turned upside down, but it certainly has been turned on its side! A decade ago, if confronted with the data matrix shown on the right of this figure, a well-trained information scientist would say "This is ridiculous. Why would you ask me to analyze a data set where you clearly have a profoundly under-determined problem? There's not enough data here to distinguish between any of the zillion hypotheses that could be consistent with this data set. And who designed these experiments, anyway? How can you measure so many features of such a few examples?" Yet, these experiments are proceeding and are making major contributions to our understanding of how gene systems interact, how to distinguish different types of cancer, and how to measure the impact of the environment on a cell. Our information scientist friend is, in some sense, correct about the relative paucity of data. (Have you ever tried to convinced a biologist holding a microarray with 45,000 spots that this is a relatively data-poor exercise? It's not fun.) However, the information scientist has missed the point about the design and analysis of these experiments. These data sets do indeed contain gold, but the experiments (as for all experiments) must be considered carefully in both the design and implementation phases in order to maximize value. This is where the authors of this book have made a contribution. They start from the premise that these experiments offer great potential, but must be performed and analyzed carefully. They set the context of traditional reductionist biology, and then go on to discuss the design, analysis, storage and interpretation of this first generation of functional genomics experiments. The writing is lively and candid, and the examples are taken from an array of applications. The authors' practical experience in dealing with this data comes through, and they intersperse practical advice with philosophical reverie. Sometimes, these two merge into important discussions such as on the role of ontologies in making sense of these data sets, or on the challenges of linking microarray results with phenotypic data pertinent to human disease.

The functional genomics revolution is here. We do not know how it will change our view of biology and medicine. They are both much more likely to become quantitative and systematic (as opposed to qualitative and reductionist). The informatics techniques required to address this revolution are not entirely clear, but this text gets us started in the right direction.

Russ Altman
Stanford University
March 2002

Microarrays for an Integrative Genomics

Isaac S. Kohane, Alvin T. Kho, and Atul J. Butte

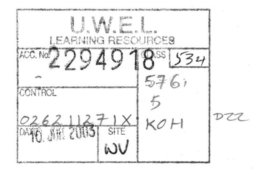
A Bradford Book
The MIT Press
Cambridge, Massachusetts
London, England

This book was set in in Computer Modern by the authors using the LATEX typesetting system and was printed and bound in the United States of America.

Library of Congress Cataloging-in-Publication Data

Kohane, Isaac S.
Microarrays for an integrative genomics /   Isaac S. Kohane, Alvin Kho, Atul J. Butte
        p.   cm.—(Computational molecular biology)
    Includes bibliographical references.
    ISBN 0-262-11271-X (hc.: alk. paper)
    1. DNA microarrays. 2. Gene expression—Analysis—Automation. 3. Bioinformatics. I. Kho, Alvin. II. Butte, Atul J. III. Title. IV. Series.
QP624.5.D726 K686 2002
572.8′6–dc21   2002022663

Contents

Foreword

The impact of microarray measurements on biology and bioinformatics has been astounding. Starting from virtually no literature a few years ago, this field has come to dominate many conferences and journals. As an example, the Intelligent Systems for Molecular Biology conference, the annual meeting of the International Society for Computational Biology held in Copenhagen in 2002, had almost 50% of its papers in the areas addressed directly or indirectly within this book. Four years ago, there were none. Bioinformatics has always been driven by the availability of data—sequential, structural, and most recently functional. The availability of sequence data brought into biology a cadre of computer scientists with special skills in string processing. The availability of structural data brought in technical experts in visualization and computational geometry. This most recent development—the availability of relatively large data sets measuring the expression of genes within cells—has helped attract yet another group of scientists—the data miners, machine learners and statisticians.

In many ways, the impact of this data on biology and informatics can be summarized in figure 1.4 of this book—the world has not quite been turned upside down, but it certainly has been turned on its side! A decade ago, if confronted with the data matrix shown on the right of this figure, a well-trained information scientist would say "This is ridiculous. Why would you ask me to analyze a data set where you clearly have a profoundly under-determined problem? There's not enough data here to distinguish between any of the zillion hypotheses that could be consistent with this data set. And who designed these experiments, anyway? How can you measure so many features of such a few examples?" Yet, these experiments are proceeding and are making major contributions to our understanding of how gene systems interact, how to distinguish different types of cancer, and how to measure the impact of the environment on a cell. Our information scientist friend is, in some sense, correct about the relative paucity of data. (Have you ever tried to convinced a biologist holding a microarray with 45,000 spots that this is a relatively data-poor exercise? It's not fun.) However, the information scientist has missed the point about the design and analysis of these experiments. These data sets do indeed contain gold, but the experiments (as for all experiments) must be considered carefully in both the design and implementation phases in order to maximize value. This is where the authors of this book have made a contribution. They start from the premise that these experiments offer great potential, but must be performed and analyzed carefully. They set the context of traditional reductionist biology, and then go on to discuss the design, analysis, storage and interpretation of this first generation of functional genomics experiments. The writing is lively and candid, and the examples are taken from an array of applications. The authors' practical experience in dealing with this data comes through, and they intersperse practical advice with philosophical reverie. Sometimes, these two merge into important discussions such as on the role of ontologies in making sense of these data sets, or on the challenges of linking microarray results with phenotypic data pertinent to human disease.

The functional genomics revolution is here. We do not know how it will change our view of biology and medicine. They are both much more likely to become quantitative and systematic (as opposed to qualitative and reductionist). The informatics techniques required to address this revolution are not entirely clear, but this text gets us started in the right direction.

Russ Altman
Stanford University
March 2002

Preface

Three years ago, when a colleague would approach us with questions about functional genomics and the informatics techniques required to leverage the data obtained from measurement techniques such as DNA microarrays, we had a standard response: "Come listen to a 1-hour presentation by one of us and you'll have a foundation for further discussions." Since then, this response has become inadequate. First, the field can hardly be summarized in even eight 1-hour lectures, and second, the growth in the number of potential collaborators has far outstripped the time available to us to make the necessary presentations.

In early 2000, Ben Reis, one of the graduates of the Children's Hospital Informatics Program, had the inspired idea of formalizing our introductory presentations into a book. We immediately agreed that this was a timely suggestion and to its credit, so did The MIT Press. In our teaching duties in several courses within the Division of Health Sciences and Technology (HST) at Harvard/MIT the range of topics within functional genomics that we were covering in formal presentations grew rapidly. Subsequently, with the inception of the Bioinformatics and Integrative Genomics training program at HST and the development of a Genomic Medicine course at HST, we felt the need for this book all the more acutely.

Organization of the text

We recognize that the readership of this book will be varied due to the intrinsically multidisciplinary nature of the functional genomics enterprise (as will be emphasized in the introductory chapter). Accordingly we outline the content of the following chapters so that readers may choose for themselves the path that suits them. Nonetheless, our intent and contention is that the current ordering of the chapters provides the most efficient way of acquiring the content of this book.

Introduction. Here we establish the motivation and the scope of this book and touch upon substantial obstacles to success in the successful application of bioinformatics to an integrative genomics. The notion of an interdisciplinary functional genomics pipeline is introduced. We also review which kind of readers might find this book worthwhile. The promise and limitations of functional genomics techniques, the nature of various kinds of genomic data, and the central role played by the discipline of bioinformatics are outlined. For those who have a limited background in biological sciences, there is a subsection on the basic minimum of molecular biology concepts that will be needed to grasp the the following chapters.

Chapter 1. Experimental Design. This chapter develops a framework for approaching the design of microarray-driven functional genomics experiments. Very little here is quantitative or mathematical. Rather the emphasis is on ways of thinking about the design of experiments and how it might impact the yield of these experiments. We address challenges that are particular to computer scientists (*e.g.*, defining a figure merit for the performance of the bioinformatics algorithms) and to biologists (*e.g.*, discarding potentially valuable data using formal decision theory because of the scale issues in massively parallel data acquisition using noisy measurement devices), respectively. In exploring the design issues we introduce the functional genomics clustering dogma, the broad machine-learning categories of supervised and unsupervised learning, and the nature of the analyses developed using these techniques.

Chapter 2. Microarray Measurements to Analyses. We lay the foundations for performing analyses of microarray data sets. This is the first of the more quantitative and mathematical chapters. We start with a discussion of the acquisition of digital data from the two most widely employed classes of microarrays. Then we consider the two most generic problems of comparing gene expression within a single microarray, *i.e.*, *intra*-array analyses, and comparing expression across microarrays, *i.e.*, *inter*-array analyses; in so doing, we introduce the fundamental concept of (dis)similarity and similarity measures and the several kinds of such measures. These measures become the building blocks for the genomic data-mining techniques described in the following chapter.

Chapter 3. Genomic Data-Mining Techniques. When gene expression is measured in more than two samples, gene expression patterns have to be analyzed using methods that consider the coordinated interactions of genes across multiple conditions. This chapter assesses the components of biomedical experiments that can be included in a data-mining investigation. We then cover the most commonly used analytic techniques, discussing the advantages and disadvantages of each technique, as well as the postanalysis process. Where appropriate, we provide pseudocode that will allow readers with some training in computer science to understand the details of the most often used and cited data-mining algorithms. The emerging field of genetic network reverse engineering is also introduced here.

Chapter 4. Bio-Ontologies, Data Models, Nomenclature. This chapter addresses possibly the least exciting but the most pressing bioinformatics need for genomic research: creating and using comprehensive annotations of gene function, storing and organizing microarray expression data, and ensuring standardized access to these data. We review current efforts to create formalized systems of description of gene function and the various kinds of "ontologies" that support these descriptions. The challenge to design "standardized data models" for the storage of microarray data is addressed and the principal contenders claiming to be this standard are reviewed. Naming schemes—nomenclatures—most applicable to gene expression studies are described. Nomenclatures, data models, and ontologies are placed in a perspective of the general problem of analyzing the results of functional genomics experiments. Tools that leverage these standardization efforts and the on-line published literature are also described.

Chapter 5. From Functional Genomics to Clinical Relevance: Getting the Phenotype Right. Here we address the process of translating the functional genomics research agenda into one of clinical relevance. We place in this perspective the value and deficiencies of electronic medical records and standardized clinical vocabularies. Although by no means comprehensive, we provide the highlights of the privacy issues (*e.g.*, the implications of the Health Insurance Portability and Accountability Act, anonymization, cryptographic identifiers, *etc.*) that are most likely to have an impact on the clinical application of genomic technologies.

Chapter 6. The Near Future. As the techniques and goals of functional genomics are in rapid flux, we engage in some short-term forecasting to guide readers planning in this time window. Microarray technologies being developed and recently released are previewed. In this context, the problem of comparing expression measurements across generations of microarray measurement platforms is appraised. More broadly, the different kinds of software required for the successful functional genomics enterprise are described. Finally, a model to meet the training needs of this new discipline is outlined.

Acknowledgments

ISK would like to thank his two co-authors: Without their belief in the worth of this collaborative enterprise and without their expertise, this book would never have been completed. Dozens of colleagues generously provided of their knowledge and resources including Hamish Fraser, Steven Greenberg, Winston Kuo, Ashish Nimgaonkar, Peter Park, Marco Ramoni, Alberto Riva, Ben Reis, Zoltan Szallasi, Peter Szolovits, and Christine Tsien. Our colleagues in industry, notably Bill Craumer, John Hart, and Bill Buffington were equally generous. Of course, all errors and misinterpretations remain the authors'. David Ruckle is due thanks for his effective transmogrification of the authors' scribbles into attractive illustrations. To Marie Boyle, ISK offers his gratitude for her unparalleled organization skills. Robert Prior would do any publisher proud for his thoughtfulness and integrity. Elaine and Leo provided a quiet and hospitable haven for ISK before some particularly tight deadlines. To Heidi, ISK is thankful for the love, understanding and encouragement that enabled this work to proceed even when sunny days, and other enticing distractions beckoned. Finally, to Judith and the late Akiva Kohane who transmitted the core values of which this effort is a small reflection, ISK will always be in their debt.

ISK's time on this project has been funded in part through the generosity of the John F. and Virginia B. Taplin Award of the Harvard-MIT Division of Health Sciences and Technology as well through funding by the NIH including N01 LM-9-3536 "Personal Internetworked Notary and Guardian" from the National Library of medicine, HL066582-01 and HL-99-24 through the Program for Genomic Applications of the National Heart Lung and Blood Institute, U24 DK058739 "NIDDK Biotechnology Center" by the National Institute of Diabetes, Digestive and Kidney Diseases, 1R21 NS41764-01 "Functional Genomic Analysis of the Developing Cerebellum" by the National Institute of Neurological Disorders and Stroke, U01 CA091429-01 "Shared Pathology Informatics Network" of the National Cancer Institute, and P01 NS 40828-01 "Gene Expression in Normal and Diseased Muscle During Development" by the National Institute of Neurological Disorders and Stroke.

ATK is deeply indebted to his co-authors for their immense patience and learning in answering and enduring his oftentimes naïve biologic queries; To them and Marie Boyle, our work-besieged CHIP administrator, for their unrelenting cheerfulness and easy camaraderie during our writing spells. ATK is very grateful to David Rowitch for funding support via the National Institutes of Health grant 1R21 NS041764-01 "Functional Genomic Analysis of the Developing Cercbellum" throughout the duration of this project. Lastly, but not in the least, he expresses his deep gratitude to his parents Kwang Khoon and Rose Kho for their ever constant love and affection.

AJB is first indebted to his co-authors for the stimulating discussions needed to prepare this book, many of which have resulted in the design and development of novel bioinformatics methods while this book was being written. In addition, without Marie Boyle, this book would never have been written or organized. AJB wishes to thank his friend and mentor, Dr. Isaac Kohane, for accepting him into the Children's Hospital Informatics Program four years ago. AJB sincerely owes his being in this field of research to Dr. Kohane's early vision and strong collaborative connections to biologists. AJB appreciates Dr. Kohane's continued support and development of his career in bioinformatics. AJB also wishes to thank his division chief, Dr. Joseph Majzoub, for his support of AJB's clinical and research environment, and Professor Peter Szolovits for his guidance and advice. AJB remembers his uncle, the late Dr. Prakash Kulkarni, who inspired him to become a pediatrician, and also thanks Dr. Simeon Taylor and Dr. Michael Quon, both of whom taught AJB everything he knows about the laboratory study of molecular biology and inspired him to enter the field of endocrinology. AJB wishes to thank his brother, Dr. Manish Butte, for being a role model with his academic pursuits and his discussions on the nature of biophysics and bioinformatics, and his parents Janardhan and Mangala Butte, for their love, support, and wisdom in exposing their children to computers at an early age. Finally, AJB wishes to thank his wife, Dr. Tarangini Deshpande, for her love, support, and her encouragement to complete this work.

During the writing of this work, AJB has been funded by and wishes to thank the Endocrine Fellows Foundation, the Genentech Center for Clinical Research and Education, the Lawson Wilkins Pediatric Endocrinology Society, the Harvard Center for Neurodegenerative Research, and the Merck-Massachusetts Institute of Technology partnership. AJB was also supported in part by the grant "Genomics of Cardiovascular Development, Adaptation, and Remodeling" funded by the National Heart Lung and Blood Institute's Program in Genomic Applications, U01 HL066582, the grant "Harvard-MIT-NEMC Research Training in Health Informatics", funded by the National Library of Medicine, 5T15 LM07092, and the grant "NIDDK Biotechnology Center", funded by the National Institute of Diabetes, Digestive and Kidney Diseases, U24 DK058739.

Microarrays for an Integrative Genomics

1 Introduction

The functional genomics "meltdown" is coming. At least that is what we fear is likely to occur with the confluence of the high expectations engendered by the Human Genome Project and the prevalence of highly uneven scholarship in the investigations made possible by comprehensive genomic measurement technologies, such as microarray-based expression profiling. Because the availability of these technologies has preceded the development of a substantive canon of appropriate analytic techniques, safe experimental design, and cautionary tales, the quality of these investigations and the manner in which they have been reported often results in the dissemination of highly preliminary and flimsy findings. The absence of widespread use of computational and biological validation procedures associated with these studies has led to many reports that will likely not be substantiated by follow-up studies. With the rapid decrease in the cost of genomic techniques, follow-up studies are likely to arise within the next few years. At that time, several previously well-publicized findings in basic biology, clinical diagnosis, clinical prognosis, and pharmacogenomic targeting will be found deficient. The inevitable reaction to these unfortunate developments will be attenuated if the discipline of functional genomics has matured enough by then to achieve the rigor required.

In this book, we have attempted to address some of the challenges required to begin this maturational process in the intersection between microarray expression technology, bioinformatics, and biomedical science. Despite the likelihood of an eventual disappointment and subsequent retrenchment of the ambitions of microarray applications, we are confident that the development of systematic approaches to this discipline will ultimately deliver on the exuberant predictions and promises made over the last 5 years. We hope that our experience (including our own generous share of mistakes), as communicated here, will help provide the reader with a framework to participate productively in attaining this goal.

1.1 The Future Is So Bright...

Let us be clear lest the above suggest that we are pessimistic about this field. There are very few disciplines within biomedical research with as much promise and excitment as functional genomics. Consider the example of the analysis of large B-cell lymphoma, a deadly malignancy of the lymphatic system, conducted by Alizadeh *et al.* [3]. In (a necessarily abbreviated) summary of their study, the gene expression of the lymphatic tissues of a cohort of patients with the diagnosis of large B-cell lymphoma was measured using DNA microarray technology. That is, thousands of genes expressed in these tissues were simultaneously measured in the respective lymphatic tissue sample of each patient.

When a clustering analysis was performed to see which patients ressembled one another the most, based on their gene expression pattern, two distinct clusters of patients were found. When the investigators examined the patients' histories it became apparent that the two clusters corresponded to two populations of patients with dramatically different mortality rates (illustrated by the survival curves in figure 4.13, section 4.11).

The implications of these two significantly distinct mortality rates are profound. First, these investigators have discovered with genomic technologies and bioinformatics analyses a new subcategory of large B-cell lymphoma, a new *diagnosis* with clinical significance. Second, they have generated a tool that provides (pending confirmation in other studies) a new *prognosis*; patients can be given much more precise estimates of their longevity. Third, it provides a new *therapeutic opportunity*; patients with an expression pattern predicting a poor response to standard therapy may be treated with different (*e.g.*, much more aggressive) chemotherapy. Fourth, it presents a new *biomedical research* opportunity; what is it about these two subpopulations that makes them so different in outcome and how can that be related to the differences in gene expression?

It is rare that a set of measurement and analytic techniques can so revolutionize biomedical research and clinical practice. It is precisely because the excitement and expectations surrounding this field are so high that we are compelled to inject a note of skepticism about the measurement techniques and the analytic methods. Without such skepticism, the scientific method cannot function and the field will not advance. Nevertheless, even as we have discovered for ourselves significant drawbacks in genomic methodologies, which we address in this book, we remain convinced that these problems only represent the transient growing pains of a new field of biomedical investigation.

Who is this book intended for? Answering this question has served as our constant compass throughout the book's writing. There are three audiences that we have had in mind.

1. *Experienced biologists with limited experience using expression microarrays, or who are concerned that their current approaches to this field are problematic.* For them, this book provides a systematic approach to using microarrays as a tool to investigate biology. Our particular goal for this audience is to realize the pitfalls and *caveats* relating to expression microarray technologies and their analysis. Also, we intend this book to provide these biologists with a firm foundation on which they

can engage in collaborations with colleagues formally trained in bioinformatics and biostatistics. We deliberately limited the amount of mathematical treatment of this material. Where it is present, the reader can skip it without significantly reducing the comprehension of the subsequent text.

2. *Experienced informaticians with limited experience analyzing microarray data.* We believe the field of functional genomics to be one of the most fruitful and rewarding for a computer scientist or other quantitatively trained scientist to enter. It provides a source of challenges, problems, and data sets that will stimulate basic methodological development while furthering important goals in the enterprise of biological discovery and the state of the art of clinical care. For this reason, we emphasize an approach that is driven by the questions of interest to these latter goals. As a result, this book will not present the fundamental computer science underlying various techniques (*e.g.*, proofs of the soundness of various machine-learning algorithms) but will instead illustrate their application to problems that are challenging investigators in functional genomics. This is not to say that the approach we have taken is not rigorous; it just eschews details on the methodologies that are available elsewhere and that we have copiously referenced.

We are convinced that the most productive research projects will be those that involve intimate collaborations with biologists. This is in contrast to the remote and *post hoc* analysis of the outcome of experiments that the bioinformatician has little to do with, but which is frequently the norm in this discipline. We intend and hope this book can serve as the basis for collaborations in which the informatician understands the goals of biology in this scientific enterprise and in which she or he can contribute to the experimental design and analysis as a first-class member of the research team.

3. *Students entering the field of Bioinformatics.* In our own classes (*e.g.*, Medical Computing 6.872 taught at MIT), we have been gratified to note the emergence of a new generation of students who are both facile in the use of computers as experimental tools and who have a broad understanding of the biological sciences. Although this text can be used as the basis of a course, it is also designed for independent study. For those students, this text is intended to serve as a rapid introduction to the fields of microarray-driven studies of functional genomics so that they can become productive researchers even while they pursue their studies. Indeed, we have had several successful collaborations with undergraduates who have used this material as a launching pad for graduate research or careers in industry.

1.2 Functional Genomics

Now that the human genome has been sequenced (or, more accurately, now that a handful of human genomes have been sequenced), we are said to be in a postgenomic era [200]. We find this term confusing ("genome, we hardly knew ye") because in our view, now that we know at least the draft outline for the genome of multiple organisms, we can begin for the first time to systematically deconstruct how the genetically programmed behavior of an organism's physiology is related to the constituent genes that make its individual version of its species' genome. In this deconstruction, several kinds of biological information are available: DNA sequence, physical maps, gene maps, gene polymorphisms, protein structure, gene expression, protein interaction effects,[1] and a vast literature in MEDLINE [129].

As we use it in this book, functional genomics refers to the overall enterprise of the deconstruction of the genome to assign biological function to genes, groups of genes, and particular gene interactions. These functions may be directly or indirectly the result of a gene's transcription. Much of functional genomics has been and will continue to be the kind of hypothesis-driven biological research pursued for the past decades.

In this book, we address a computationally intensive branch of functional genomics that has emerged as a result of the practical implementation of technologies to assess gene expression thousands of genes at a time. The ability to comperhensively measure expression affords an opportunity to reduce our dependence on *a priori* knowledge (or biases) and allow the organism to point us in potentially fruitful directions of investigation. That is, we describe a hypothesis-generating effort which, if carefully crafted, can then lead to a highly productive set of investigations using more conventional hypothesis-driven research. In this we have been inspired by the work of Arkin *et al.* [12], as have others. Starting from the raw time-series measurements of the substrates of glycolysis (see figure 1.1), Arkin *et al.* were able to computationally reconstruct the glycolytic pathway (figure 1.2). This reverse engineering of the metabolic pathway without prior knowledge is in contrast to the decades of exacting hypothesis-driven elucidation of this pathway by biochemists. It turns out that, as we will discuss later, in experimental design (section 2.1.3), this metaphor is flawed but it remains an icon for one of our major goals: to use bioinformatics applied to functional genomics data to create and re-create the kind of biological pathway charts that are common in most basic biology laboratories.

---

[1] For those of you who are unfamiliar with what these are, we touch on defining these terms in section 1.5 and in the glossary.

Furthermore, we note that genomic data can be fruitfully exploited even without the assignment of function: a prognostic test for rejection of transplanted kidneys based on the expression level of three genes is useful even if the function of those three genes is poorly known or not known at all.

1.2.1 Informatics and advances in enabling technology

Gene expression detection microarrays are notable not because they can uniquely measure gene expression. There certainly have been many technologies that have allowed for the quantitative or semiquantitative measurement of expression for well over two decades. What distinguishes gene expression detection microarrays is that they are able to measure tens of thousands of genes at a time and it is this *quantitative* change in the scale of gene measurement that has led to a *qualitative* change in our ability to understand regulatory processes occurring at the cellular level. Figure 1.3 provides perhaps the best motivation for the application of information sciences to the functional genomics enterprise. Since DNA sequencing was invented 25 years ago, the number of gene sequences deposited in international repositories, such as GenBank, has grown exponentially, culminating with the entire human genome being sequenced in 2001. Distinguished from this, the amount of *knowledge* about these genes (as measured by the proxy of the number of papers published in biomedicine) has also been growing exponentially, but at a *much* slower rate. As shown, the number of GenBank entries has fast outstripped the growth of MEDLINE. As such, it serves as a proxy for the large gap that has just opened up between our knowledge of the functioning of the genome and raw genomic data. And GenBank entries are just a fraction of the various kinds of data, listed above, generated as part of our investigation of the genome. This volume of data must somehow be sifted and linked to the biological phenomena of interest. Doing so exhaustively, reliably, and reproducibly is a plausible strategy only with the application of algorithmic implementations on computers. This has led to an unprecedented demand for investigators with the knowledge of successful manipulation and analysis of large data sets. These skills may come from education and training in computational physics, chemical engineering, operations research, or financial modeling, but once they are applied to the domain of functional genomics, they can be collectively described as belonging to the domain of bioinformatics.

One reason why the number of known sequences is growing so much faster is the discovery and use of many automated techniques, such as automated sequencers and shotgun sequencing methods. Until the recent advent of gene expression microarrays, we did not have a similar technique to automate the acquisition of knowledge about these genes' behavior in cellular physiology. The past 5 years

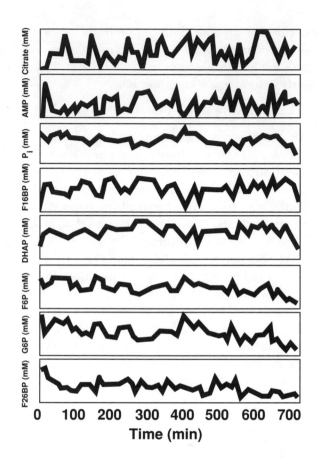

Figure 1.1
Time-series data from anaerobic metabolism. The time courses of measured concentration of the
small molecule inputs, adenine monophosphate (AMP), a source of chemical energy to catalyze
reactions) and citrate (a substrate), in the experiments, with the responses of the concentrations
of phosphate (P_i, an inorganic ion) and of the substrates fructose-1,6-biphosphate (F16BP),
dihydroxy acetone phosphate (DHAP), fructose-6-phosphate (F6P), glucose-6-phosphate (G6P),
and fructose-2,6-biphosphate (F26BP). (Derived from Arkin *et al.* [12].)

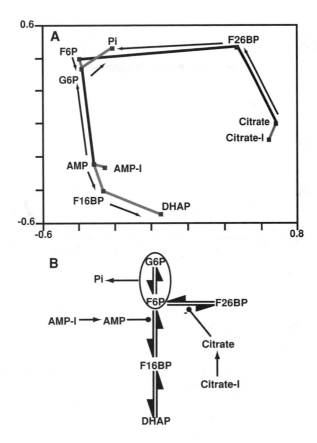

Figure 1.2
Glycolytic pathway reconstructed *ab initio* from time-series data. **A,** The two-dimensional projection of the correlation metric construction (CMC), defined by Arkin *et al.*, for the time series shown in figure 1.1. Each point represents the time series of a given species. The closer two points are, the higher the correlation between the respective time series. Black (gray) lines indicate negative (positive) correlation between the respective species. Arrows indicate temporal ordering among species based on the lagged correlations between the their time series. **B,** The predicted reaction pathway-derived CMC diagram. Its correspondence to the known mechanism of glycolysis is high. (Derived from Arkin *et al.* [12].)

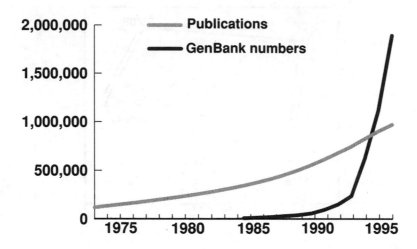

Figure 1.3
Relative growth of MEDLINE and GenBank. The industrialization of genomic data acquisition has created a growing gap between knowledge—of which MEDLINE publications are a proxy—and the information we have gathered about the genome. Information science applied to this information (*i.e.*, bioinformatics) is one of the pillars of an international strategy to overcome this knowledge gap. Cumulative growth of molecular biology and genetics literature (light gray) is compared here with DNA sequences (dark gray). Articles in the G5 (molecular biology and genetics) subset of MEDLINE are plotted alongside DNA sequence records in GenBank over the same time period. (Derived from Ermolaeva *et al.* [65].)

have seen an incredible confluence of disparate technologies, such as robotics, flo-rescence detection, photolithography, and the Human Genome Project,[2], so that today, biologists can use RNA expression microarray detection technologies to ob-tain near-comprehensive expression data for individual cells, tissues, or organs in various states. With currently available commercial tools, a single experiment using RNA expression detection microarrays can now provide systematic quantitative in-formation on the expression of 60,000 unique RNAs within cells in any given state. Complementary DNA (cDNA) and oligonucleotide microarray technology[3] cannot only be used to determine the abundance of RNA transcripts. By virtue of their broad reach, these measurement platforms permit a large number of exhaustive comparisons: of transcriptional activity across different tissues in the same organ-ism, across neighboring cells of different types in the same tissue, across groups of patients with and without a particular disease or with two different diseases. These platforms can also be used to analyze complex systems, such as traits with multigenic origins or those linked to the environment. They can be used in time se-ries to measure how a particular intervention may start a transcriptional program, *i.e.*, change the expression of large numbers of genes in a reproducible pattern de-termined by inherent genetic regulatory networks. With sufficient data, they can be even used to provide insight into the underlying mechanisms of these genetic regulatory networks.

Nonetheless, the tools to extract *knowledge* from *data* collected from all of these types of experiments are still in their infancy, and novel tools are still needed to sift through the enormous databases of simultaneous RNA expression to find the true nuggets of related function. The application of techniques of information science, computer science, and biostatistics[4] to the challenge of knowledge acquisition from genomic data is commonly known as *bioinformatics*. This appellation applies to the quantitative and computational analysis of all forms of genomic data, including gene sequence, protein interactions, protein folding, and any observable or measurable phenomenon of interest to the biomedical researcher. The breadth of this commonly used definition of bioinformatics risks relegating it to the dustbin of labels too general to be useful of which artificial intelligence, knowledge management, and systems analysis are only among the more recent. In this book our intent is to be

---

[2] Using the common shorthand title for the international public and private effort to determine the sequence of bases in the human genome.

[3] The characteristics of the various microarray technologies are addressed in chapter 3.

[4] The label applied often seems to be determined more by the training background of the labeler rather than any fundamental characteristic of the analytic technique.

sufficiently specific about the bioinformatics techniques employed that the matter
of a sufficiently broad and yet specific definition of bioinformatics is moot.

Over the past 6 years, several approaches have been developed to analyze microarray-
generated RNA expression data sets. The central hypothesis (or hope) of these
methods is that, with improved techniques in bioinformatics, one can analyze larger
data sets of measurements from RNA expression detection microarrays to discover
the "true" biological functional pathways in gene regulation, and develop more
definitive, sensitive, and specific diagnostic and prognostic characteristics of disease.
However, this is only one of many important areas of bioinformatics addressed in
this book. Particularly because we are still in the immediate aftermath of the Human Genome Project, many of the basic naming and data management practices of
functional genomics remain in flux and are active areas of bioinformatics development. Although this activity is quite distinct from the analytic efforts touched on
above, it currently consumes perhaps the largest proportion of the bioinformatics
community because its resolution is urgent and a *sine qua non* for the success of
any of the analytic efforts. After all, if we cannot reliably name the same gene in
identical fashion across experiments, if we cannot reliably retrieve expression data
from all the microarray experiments of interest, if we cannot readily access the
meaning and function of genes determined by thousands of researchers, then the
whole enterprise of functional genomics will be crippled if not intractable.

There is a related discipline to bioinformatics—*clinical informatics*—which refers
to the application of information science to various aspects of clinical care. Although
clinical informatics is not addressed in this book in detail, in chapter 6 we describe
many of the problems that have dogged clinical informaticians and that will confront
bioinformaticians as they attempt to bring their basic science findings to clinical
relevance.

1.2.2 Why do we need new techniques?

A first look at a typical genomic study might cause a quantitatively trained scientist
or even a biologically trained scientist to ask the following, quite legitimate question:
Why is this field not amenable to standard biostatistical techniques? After all,
we are trying to understand the relationship between multiple variables and the
mechanisms that the relationships reveal. And there has been a long history of
the development of biostatistical techniques to analyze large studies with large
numbers of cases with many variables to elucidate precisely this kind of question.
Specifically, these studies ask questions such as: What risk factors are associated
with heart disease? Does smoking cause disease? What is the difference in survival
between a group treated with one chemotherapeutic drug versus another? On

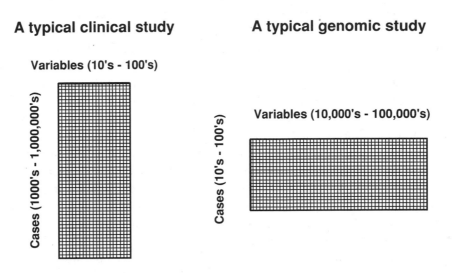

Figure 1.4
A major difference between classic clinical studies and microarray analyses. The high
dimensionality of genomic data in contrast to the relatively small number of samples typically
obtained results in a highly underdetermined system.

the surface these questions seem similar to many of those posed regarding genetic
risk factors for acute and chronic disease. Yet a review of the bioinformatics and
functional genomics literature over the past 3 years reveals that most of the analyses
have been performed using techniques borrowed from the computational sciences
and machine-learning communities in particular. Why is this? There are several
reasons, including academic parochialism, but perhaps the most substantive one is
the essentially underdetermined nature of genomic data sets as described below. If
we examine figure 1.4, we see sketched out the fundamental difference between a
typical clinical study and a typical genomic study. A high-quality clinical study
will involve thousands to tens of thousands of cases, such as in the Nurses' Health
Study [18] or the Framingham Heart Study [55] over which tens or even hundreds
of variables are measured. In contrast, in a typical genomic study, there are only
tens or, exceptionally, hundreds of cases, but thousands of measured variables.

Initially, the low number of cases in functional genomic investigations may have
been due to the high cost of the microarrays (on the order of several thousand
dollars per microarray) but increasingly the scarcity of cases in a typical functional
genomics study will relate to the scarcity of appropriate biological samples. As

these experiments involve the measurement of gene expression, a particular tissue
has to be obtained under the right conditions. This is in distinction to genomic
DNA samples where most blood samples will suffice. Especially in human populations, suitable tissue samples are relatively rare.[5] Yet even though there are
only tens of cases, each case involves the measurements of tens of thousands of
variables corresponding to the expression of tens of thousands of genes measurable
with microarray technology. The result of the large number of variables compared
to the number of cases is that we have highly *underdetermined* systems. That is,
we are making measurements of very high dimensionality (on the order of tens of
thousands) but we are only providing a small number of cases to explore this high-
dimensional space. Another way to say this is that there are many, many ways in
which the variables being measured could be interrelated mechanistically, based on
the relatively small number of observations. Due to this high dimensionality and
the underdetermined nature of these systems, standard biostatistical techniques do
not hold up well because many of the assumptions that underlie these conventional
biostatistical techniques do not hold.[6] We often provide the following analogy. To
solve a linear equation of one variable (*e.g.*, $4x = 5$) we only need one equation
to find the value of the variable. To solve a linear equation of two variables (*e.g.*,
$y = 4x + b$), two equations are required. If we have tens of thousands of variables,
but only hundreds of equations, then there will be thousands of potentially valid
solutions. This is the essence of what constitutes an underdetermined system. In
this context, we must use techniques that can maximally inform us of the relationships between variables of interest (and find out which ones are of interest) despite
the underdetermined nature of the data sets. High-dimensionality data sets have
been well-known to the "machine-learning" community of computer scientists in
applications such as automated recognition of human faces, so it is not surprising
that many of the techniques developed by that community have found their way
into the functional genomics enterprise.

---

[5] See chapter 2 for a discussion of which tissues are appropriate for particular experiments.

[6] Although this holds true for most of the biostatistical techniques biologists will have learned
in graduate school, in fairness there has been quite a lot of research by statisticians on the analysis
of underdetermined systems of high dimensionality. Their work has just not found its way into
mainstream biomedical study until recently.

diseases like hypertension, it is not clear what the functionally relevant tissue is. A successful pipeline involves collaboration with a source of tissue, such as a surgical team, a laboratory with biologically interesting animals, or a laboratory with cell lines of interest.

- *Right conditions.* Even if the appropriate tissue is selected from the organism of interest, the conditions under which the tissue is obtained (*e.g.*, number of hours post mortem) can determine whether or not the investigation is successful. An insulin-sensitive tissue such as skeletal muscle will have a different characteristic metabolic and expression profile depending on the glucose and insulin concentrations prior to the extraction of RNA. The time of day will influence the expression of genes in all tissues which have endogenous circadian rhythms or that have processes that can be entrained by physiological clocks. Awareness of these issues and cooperation from a surgeon, pathologist, or technician responsible for obtaining the tissue is therefore an essential component to the success of the functional genomic pipeline.

- *Extracting RNA, hybridizing to microarray, and scanning.* Each of these steps in this "wet" component of a functional genomics pipeline is susceptible to operator error and is a potential source of poor or noisy measurements. The RNA extracted may be of poor quality, the hybridization conditions may vary (*e.g.*, the room temperature), and the settings of the scanner that produces the digital image of the microarray may vary from one scan to another. Industrialization and standardization of this component has been the focus of the more successful and high-quality functional genomics efforts using expression microarrays.

- *Functional clustering.* This "dry" component of the pipeline is often thought to be what bioinformatics is about. And in fact, it may be at this stage that the algorithmic analysis of an expression profiling study to detect biologically or clinically meaningful patterns or associations is the only time a bioinformatician is involved. We will argue throughout this book that a successful functional genomics pipeline involves the bioinformatician at every step.

- *Computational validation.* As will be elaborated in this book, there are many reasons to perform bioinformatics analyses on functional genomics data sets, and many methods can be used. One unique problem with these types of data sets is that they are "short and wide," meaning that many characteristics are measured on relatively few samples. For example, current microarrays offer the quantitation of up to 60,000 expressed sequence tags (ESTs) in any given sample, but current costs may limit a single experiment to 10 to 100 samples. Because of this problem, these

1.3 Missing the Forest for the Dendrograms, or One Aspect of Integrative Genomics

In the first 2 years of significant publications regarding the large scale application of microarray technologies, numerous special purpose or adapted machine-learning algorithms were described in the literature. Self-organizing maps [175], dendrograms [27, 63, 76], K-means clusters [101], support vector machines [33, 72], neural networks [107], and several other methodologies (borrowed largely from the machine-learning community of computer science) have been employed. Most of these have worked reasonably well for the purposes described in the papers. It is one of the central contentions of this book, however, that the choice of a particular clustering or classification methodology is secondary to proper experimental design and full knowledge of the properties and limitations of massively parallel expression analysis in general and those of the specific microarray technologies employed. This contention does not constitute an overly fastidious approach to functional genomics but the insight from (often expensive) experience gleaned through our own mistakes and those of our colleagues and the reported investigations. The bioinformatics technique centric approach has been evident in our own collaborations. Often, at the outset of a new investigation, our collaborators from both the computational community and the biological community immediately wish to address the questions of which are the appropriate clustering techniques and which one is "the best." While we certainly recognize that the choice of clustering or classification technique is important (and we devote chapter 4 to this matter), we are firmly convinced that it is only one part of a well-designed pipeline that defines a successful exploration of biology and medicine using microarray technology. This pipeline is diagrammed in figure 1.5 on page 17. We refer to this functional genomics pipeline throughout the book. We discuss at length the practicalities of assembling this pipeline in subsequent chapters, but a few characteristics of this pipeline bear mentioning here:

- *Selection of the right tissue.* Experiments in functional genomics require selection of the functionally relevant tissue or cell type. In certain experiments, like those using blood and solid cancers, the functionally relevant tissue is clear. In other analyses, the functionally relevant tissue is not so easily ascertainable or acquirable. For example, the clinical phenotype seen in type 2 diabetes mellitus, or insulin resistance, involves the coordinated physiological dysfunction of several organs and cell types, including liver, muscle, and fat cells. Schizophrenia involves a higher-order brain dysfunction, but brain cells are not easily accessible in humans. For some common data sets are essentially underdetermined, as described on page 10, meaning that there are many correct ways to mathematically describe the clusters and genetic regulatory networks contained within them. Thus, some computational validation is required immediately after the bioinformatics analysis so that computationally sound but biologically spurious or improbable hypotheses are screened out.

The principal motivation for the screening out of spurious or improbable hypotheses is the efforts that follow. Each hypothesis generated that passes this step may need to be validated in a biological laboratory. Some biological laboratories may wish (and may have the resources) to pursue many hypotheses and can tolerate the eventual refutation of large numbers of false-positive hypotheses. Other biological laboratories may only be able to validate a few. Thus, a proper bioinformatics analysis includes a computational validation. An ideal computational validation does not merely provide a yes or no answer as to the potential validity of a hypothesis, but instead provides a continuum of validation, or a receiver-operating characteristic curve. With such a curve, the biologist can select the desired point of sensitivity or specificity and true and false negatives and positives (see sections 2.1.4 and 4.12.3).

- *Biological validation.* Most biological questions will not be answered using microarrays. Instead, the most likely outcome from a functional genomics analysis is the next biological question to ask. As hypotheses are generated from bioinformatics analyses, biological validation is crucial to verify these hypotheses. This verification may include, for instance, making sure a particular set of genes is truly expressed at the proper time and place as hypothesized, using conventional biological techniques such as Northern blotting and *in situ* hybridization.

- *The multidisciplinary team.* In most settings, all of these steps, from acquisition of source material, to microarray construction, to bioinformatics analysis, to biological verification, cannot be performed by a single group or laboratory. A successful functional genomics pipeline brings together resources from many disciplines and of varied backgrounds. Two anecdotes serve to illustrate the value of this multidisciplinary approach.

We were in the process of analyzing a large number of microarray expression data obtained from skeletal muscle for some colleagues interested in muscular dystrophy—a class of genetic diseases of muscle. They were gratified when our clustering analyses found interactions between transcriptional factors and contractile proteins that they had discovered just months before using conventional molecular-biological techniques as well as several new but plausible interactions. However, because the clustering analyses were exhaustive, they also identified several hormonal interactions that were not of primary interest to these neuromuscular specialists. Using annotation tools linking the microarray data to several national databases, it became quickly apparent that these hormonal interactions were thought to be exclusive to adipocytes (cells constituting the principal component of fatty tissue) but we had just found suggestive evidence to the contrary. The multidisciplinary nature of our effort allowed the formulation of well-posed questions directly related to the interests of the biological investigators and yet kept us open to important hypotheses generated from the data.

We are participating in a study of the functional genomics of the developing brain using mouse models. We had computed, using approximately two dozen microarray data sets produced by our collaborators—developmental biology researchers—a list of approximately 100 genes that appeared to be involved in the development of a specific region of the brain. Our collaborators were in the process of selecting a subset of these for biological validation but we were worried about the outcome of the validation because the data had been derived from entire portions of the brain. whereas the process the developmental biologists were interested in only occurred in a minute component of the brain. It seemed probable that many of the 100 genes were not specific to the processes we were studying. Given only the expression data from the microarrays, none of the bioinformatics techniques were able to further refine or hone the list of 100. Fortunately, the developmental biologists provided us with the following insight. They knew of one gene g that was expressed in the tiny area of the brain that they were studying and they had determined empirically that it was expressed in no other part of the brain. They suggested that we find all those genes in the list of 100 that behaved very closely to g. We found 10 such genes and our collaborators went on to successfully validate 8 of them using the techniques of conventional hypothesis-driven molecular biology. If we had not drawn on the multidisciplinary capabilities of our team for that small but crucial biological insight, then we would have been stuck with a large list of nonspecific genes of little relevance to the questions originally posed by the developmental biologists.

When the initial design of the multidisciplinary functional genomics pipeline is given short shrift and the fundamental limitations of expression microarray technologies misunderstood, the enterprise of functional genomics appears to approximate the "fishing expedition" that has been the oft-stated concern of traditional biologists regarding this nascent discipline. Consequently, even if a particular investigator participates in only a fraction of the pipeline, understanding the safe

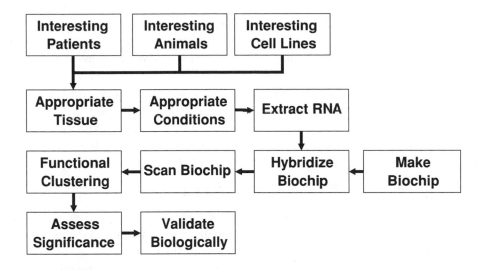

Figure 1.5
An archetypal functional genomics pipeline. Shown is a simplified view of a functional genomics
pipeline solely involved in expression microarray experiments. Note the interdigitation of "wet"
and "dry" components requiring close multidisciplinary collaboration and some creative
consideration of the value of the individual contributions in this pipeline for a particular
experiment and publication.

design of an entire functional genomics pipeline can maximize the yield of these experiments, or at the very least produce convincing and reproducible negative results. It is the intent of this book to point the way to investigations that provide such an understanding.

Because of the dramatically different backgrounds (at least at present) of the different contributors to the functional genomics pipeline, its social dynamics may be challenging, as described below. We pay attention to these dynamics because one aspect of an *integrative genomics* is the integration of disciplines and experts (the other side will be described before the end of this chapter).

1.3.1   Sociology of a functional genomics pipeline

Given the requirement for a multiinvestigator, multi-institutional effort, several pragmatic decisions and realities must be confronted. We describe these in the context of academic collaborations but analogies to the commercial world are obvious. The first question in academic collaborations around the functional genomics pipeline often is: Who will get the credit for the work and for the discoveries that ensue from a particular functional genomics investigation? More concretely, who will get first and senior authorships in the publications that report on the discoveries obtained from this functional genomics discovery pipeline? Will it be the surgeons who obtained the tissue, the bioinformaticians who performed the cluster analysis, the biologist who runs the microarray facility, or the clinician who obtained the phenotypic characterization of the patient from whom the tissue was obtained?

Resolving this issue is nontrivial because of some fundamental human and cultural considerations of the various types of investigators. Within each discipline, an investigator will tend to see those outside his or her discipline as performing more of a utility function rather than making a significant intellectual and creative contribution to the research process. For example, a molecular biologist might view the bioinformatician as providing cookbook analytic algorithms to winnow out the relevant findings from a biological system that they have spent time, energy, and creativity in developing. Conversely, the bioinformatician might view the biologist as a laborer plodding along in the murky swamps of biological experimentation on which the bioinformatician can then shed light by virtue of his or her insights into the general principles of automated inference, clustering, and classification.

For those of us who have labored on both sides of this cultural divide, it is quite clear that there are plodders and creative geniuses in both fields. Successful collaborations in this arena require thoughtful recognition of relevant contributions prior to initiating a long-term collaboration around this strategy. In our own collaborations, we have adopted the following heuristics: If the bioinformatics techniques used are not novel and the interpretation is rote, then the biologists and clinicians usually assume first and senior authorships. If the experimental design is innovative and informed by the nature of the bioinformatics analyses, and the computational techniques have to be developed or customized, then the bioinformaticians usually take first and senior authorships. Often, the major contributions are mixed, as when the bioinformatics analyses are novel and the biological validation steps are creative and arduous, in which case the authorships are split accordingly. In the end, however trite as it sounds, it is true that nothing substitutes for the collegiality and good will that arise from mutual respect for different skills and contributions.

1.4   Functional Genomics, Not Genetics

We have noticed, among some of our biological collaborators, a tendency to view the massively parallel methods of functional genomics as a highly efficient large-scale application of methods that they have already applied. For instance, gene expression profiling and polymerase chain reaction (PCR) are all methods that have been used by molecular biologists for decades. What we hope the reader will obtain from this book is an appreciation of how the near-comprehensive (and soon to be truly comprehensive) nature of the functional genomics approach as permitted by expression microarrays changes qualitatively and fundamentally the nature of biological investigation. Before our potential readers with biological backgrounds become offended by or disgruntled with this assertion, let us assure them that we present an equivalent critique for the purely computationally oriented bioinformatists and genomicists in the following section. Functional genomics is not, as some have portrayed it, a hypothesis-free fishing expedition, nor is it, even more charitably, only a hypothesis-generating enterprise requiring subsequent biological validation. It is fundamentally different in that it permits the posing of large questions that are grounded in an essential biological understanding of a particular domain. Unlike the questions posed in "traditional" genetics or molecular biology, these questions have less stringent requirements for prior supposition or claims of the role of a particular gene or metabolite in a biological process. An example of some of the broad questions that can be asked are:

- Which of all the known genes have regulatory mechanisms that appear to be similar to those regulated by the sonic hedgehog transcription factor in the cerebellum?

- Given the effect of 5000 drugs on various cancer cell lines, which gene singly is the most predictive of the responsiveness of the cell line to any chemotherapeutic agent?

- Given a known clinical distinction, such as that between acute lymphocytic leukemia and acute myelogenous leukemia, what is the minimal set of genes that can most reliably distinguish these two diseases?

- Is there a group of genes that can serve to distinguish the outcomes of patients with large B-cell lymphoma that are otherwise clinically indistinguishable on presentation?

- What distinguishes the signaling pathways of two of the substrates of the insulin receptor?

These questions are important biologically and clinically, and yet they can only be posed reasonably if they involve a comprehensive view of genomic regulation and involve the use of computational methods that can efficiently sift through the vast quantities of genomic data generated by the experiments required to answer these questions. Another way to consider functional genomics is to view it as serving as a filtered funnel through which these broad questions can be strained. The residue that remains is high-yield, detailed, and contains particular biological questions that are answerable by more traditional genetic or molecular biology techniques. This is illustrated in figure 1.6 below. The utility of this metaphor is as follows. The universe of possible participants in any given biological regulatory mechanism is finite but very large. Even with the most comprehensive in-depth expertise, a biologist may find be surprised about insights obtained through data mining of expression patterns, human genome sequences, and often from data obtained from other species. Without a genomic approach to guide her experiments, this biologist may expend several months in false leads or alternatively miss an important component to the system under study. Similarly, if the biologist is looking at a set of genes that are thought to be predictive of a given clinical condition, such as transplant rejection or cardiac disease, without the comprehensive view brought by functional genomics, elements of the diagnostic or prognostic procedure, such as the concentration of a gene transcript or a protein, may be omitted with a concomitant decrease in the sensitivity and specificity of the prognosis or diagnosis.

1.4.1 *In silico* analysis will never substitute for *in vitro* and *in vivo*

There is little doubt that one of the tremendous accomplishments of the Human Genome Project is that it has enabled a rigorous computational approach to identi-

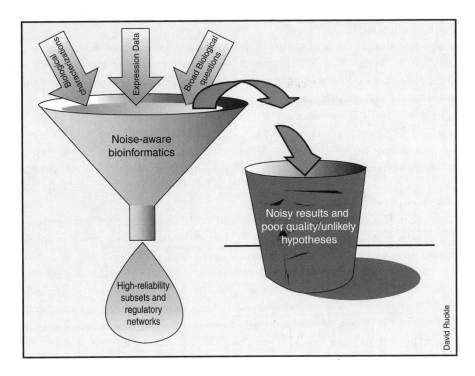

Figure 1.6
The functional genomics investigation as a funnel for traditional biological investigations. Broad questions and comprehensive data are the mix in which bioinformatics techniques are filtered to separate high-yield hypotheses or candidate genes from spurious findings and poor-quality hypotheses.

fying many questions of interest to the biological and clinical community at large.[7] However, the danger of a computational triumphalism is that its makes several dubious assumptions. The first is genetic reductionism: At an abstract level most bioinformaticians understand that a particular physiology or pathophysiology is the product of the genetic program created by the genome of an organism and its interaction with the environment throughout its development and senescence. In practice, however, a computationally oriented investigator often assumes that all regulation can be inferred from DNA sequence, based solely on the syntax of its sequence elements. That is, it is assumed that it is predictable whether a nucleotide change in a sequence will result in a different physiology. We refer to this as "sequence-level reductionism."

The second dubious assumption is the computability of complex biochemical phenomena. One of the most venerable branches of bioinformatics involves modeling molecular interactions such as the thermodynamics of protein folding, and protein-protein and protein-nucleic acid interactions. As yet, all the combined efforts and expertise of bioinformaticians have been unable to provide a thermodynamically sound folding pattern of a protein in the heterogeneous solvent environment of a cell for even as long as one microsecond. Furthermore, studies by computer scientists over the last 10 years [23, 114] suggest that the protein-folding problem is "NP hard." That is, the computational challenge belongs to a class of problems that are believed to be computationally intractable. Therefore it seems overly ambitious to imagine that within the next decade we will be able to generate robust predictive models that are able to accurately predict the interactions of thousands or millions of heterogeneous molecules and the ways in which they modulate the transcription of RNA and the translation of messenger RNA (mRNA) into protein and the subsequent functions of these proteins. We refer to this ambition as "interactional reductionism."

This is not to say that models have no useful role in molecular biology or bioinformatics. On the contrary, they are extremely useful to embody what we currently believe we know about biological systems. Where the predictive capabilities of these systems break down points to where we should guide further research. Also, the educational value of such models cannot be underestimated.

The final questionable assumption is the closed-world hypothesis. Both sequence level reductionism and interactional reductionism are predicated upon the availability of a reliable and predictive and complete mechanistic model. That is, if a

---

[7]Although even the most basic of the original conclusions, stated in the publications heralding the completion of the draft of the human genome, the order of magnitude of the number of genes in the human genome, now seems to be again in contention.

fertilized ovum can follow the genetic program to create a full human being after 9 months, then surely a computer program should be able to follow the same genetic code to deterministically infer all the physiological events that are determined by the genetic code. Indeed, there have been several efforts, such as the E-cell effort of Tomohita *et al.* [179], which aimed to provide robust models of cellular function based on the known regulatory behavior of cellular systems. Although such models have important utility, our knowledge of all the pertinent parameters for these models appears grossly incomplete today. These parameters are required to describe intracellular processes, intercellular processes, and the unimaginably large repertory of possible environmental interactions with both sets of processes. This incompleteness, and the lack of knowledge of where the boundaries are between the complete and the incomplete, imply that these models will have behaviors that may diverge substantially and unpredictably from those that actually occur.

These caricatured positions of the traditional molecular biologists and the computational biologists are, of course, overdrawn. When prompted, most of these investigators will articulate the fullness of the complexities of the analytic tasks of functional genomics. In the conduct of their research or even in the discussions within their publications, these same investigators will nonetheless often retreat to the simplifications and assumptions described above. This may be because they take for granted that their colleagues and readers understand these simplifications, but such unstated assumptions can often misdirect novices in this discipline.

What we argue for, and hope that this book communicates, is the necessity for a rapid generate-and-test paradigm that cross cuts repeatedly across the disciplines of genetics, computational biology, and molecular biology. Operationally, this means that bioinformatics tools can be used to guide the investigations of an experimental biologist investigating a particular biological system or disease process. But for even the smallest assumption, rather than relying on the statistical association or predicted behavior of a system, empirical evidence has to be developed to support these. It is only in this incremental accretion of evidence that the discipline of functional genomics can become a science.

Why microarrays? Why focus on microarrays? After all, there are many other ways to impute function to genes. Using only genomic DNA easily obtainable from a peripheral blood sample or a buccal smear, genetic epidemiologists can conduct association studies using microsatellite markers or polymorphisms [43] to associate prognoses, diagnoses, and even biological function with a particular gene [67, 126]. And then there are the more conventional genetic techniques of transgenic and misexpression whole-organism models of the function of various genes. Even more recently, the feasibility studies for proteomic assays suggest that in the future we will directly be able to assess changes in protein concentration at the cellular level.

In contrast to linkage and association studies, microarray studies are designed in principle to measure *directly* the activity of the genes involved in a particular mechanism or system rather than their association with a particular biological or clinical feature. An association-linkage-genetic epidemiology study relies upon a long indirect probabilistic causal chain: that a change in DNA sequence results in a change in gene regulation or protein structure, resulting in a change in cellular physiology measurable as a change in a whole-organism profile (*e.g.*, the human phenotype). For some changes in genomic sequence, particularly in the instances of multigenic regulation, the effects may be so small that any conceivable population study may not be able to detect them.[8] Also, the cost of screening the genome of sufficiently large populations to achieve adequate statistical power has prohibited all but the most focused association studies (although this is likely to change). Unlike the current state of art and engineering of large-scale proteomic assay systems, gene microarrays are currently affordable and within many applications have acceptable reproducibility and accuracy.

Another aspect of an integrative genomics Notwithstanding these apparent advantages of expression microarray studies, as we discuss in sections 1.5.3 and 1.5.2, there are several kinds of information that we are missing by not including such measurements. Our decision to restrict the scope of this book to the exploration of functional genomics and genomic medicine from the perspective of microarray technology is then largely a pragmatic one. Expression microarrays are sufficiently well engineered and cost-effective to allow thousands of researchers to productively employ them to drive their investigations. If, in the future, as we expect, massively parallel measurements of individual proteins becomes cost-effective, large-scale, and highly reproducible, then we will certainly expand the analysis to address these methodologies. The same will be true when high-resolution (*i.e.*, every kilobase) genome-wide scans of hundreds of individuals will become economically feasible for most clinical research studies. Current estimates have these technologies available on the genomic and population scale within 5 years. A well-prepared genomic investigator will have prepared the pipeline to take advantage of all these measurement technologies.

---

[8]That is, there would be insufficient numbers of individuals with the necessary constellation of phenotypes across the entire human population. At the same time we recognize that there are several diseases, such as sickle cell anemia, where the change of one base in the hemoglobin gene results in a severe and unsubtle disease phenotype.

From the computational perspective, the measurement of any analyte, whether it be an inorganic constituent of serum, an RNA transcript, or a protein, are all simply point measurements of variables corresponding to the total state of the cell. Likewise, all clinical measurements (*e.g.*, height, blood pressure) and history (*e.g.*, age of menarche, time of cancer diagnosis) are point measurements corresponding to the total state of the organism (*e.g.*, human). It is only in the important details of the quality and meaning of these measurements that they differ. This is said both with tongue planted firmly in cheek and in all seriousness.

And indeed, it is this other aspect of an *integrative genomics* to include as many modes of data measurement as are available. Each mode reflects another aspect of cellular and organismal physiology each with its own set of specificities and sensitivities with respect to a phenomenon or process of interest. The role of bioinformatics in an integrative genomics is to both provide the glue bringing together all kinds of genomic and phenotypic data and the means to extract knowledge (or at least high-yield hypotheses for subsequent testing) from them in an efficient, large-scale, and timely fashion. It is also the inspiration for the title of this book.

1.5   Basic Biology

This section is meant solely for the quantitatively trained scientist who knows essentially none of the biology developed over the last three decades. For those of you who need a significant refresher in molecular biology we list several good introductory texts and on-line resources (table 1.1). So we start from the beginning. In almost all cells making up a living organism, there is believed to be an identical set of codes describing the genes and their regulation. This code is encoded as one or more strands of the deoxyribonucleic acid molecule: DNA. That is the same in almost every cell in the human body.[9] For instance, a liver cell and a brain cell have the same DNA content and code in their nucleus. What distinguishes these cells from one another is that portion of their DNA that is transcribed and translated into protein, as described below.

The entire complement of DNA molecules of each organism is also known as its *genome*. The overall function of the genome is to drive the generation of molecules, mostly proteins, that will regulate the metabolism of a cell and its response to the environment, as well as provide structural integrity.

---

[9]There are several exceptions to this rule, such as mature red blood cells, which lack nuclei and therefore the organism's genome, and gametes (spermatozoa or ova), which have half the usual complement of DNA. However, for the purposes of this overview, the above generalization will suffice.

- *Genomes.*
  T. A. Brown & Austen Brown. New York, Wiley-Liss, 1999.

- *Human Molecular Genetics*, 2nd edition.
  Tom Strachan & Andrew Read. New York, Wiley-Liss, 1999.

- *Primer on molecular genetics*
  `http://www.bis.med.jhmi.edu/Dan/DOE/intro.html`.

- *Primer on genomics with a commercial flavor*
  `http://www.biospace.com/articles/genomics.primer.cfm`

- *Introductory biology course at MIT (7.01) hypertext book*
  `http://esg-www.mit.edu:8001/esgbio/701intro.html`

Table 1.1
Molecular biology primers

Structure of DNA   Each molecule of DNA may be viewed as a pair of chains of the nucleotides adenine (A), thymine (T), cytosine (C), and guanine (G). Moreover, each chain has a polarity, from $5'$ (head) to $3'$ (tail). The two strands join in opposing polarity ($5'$ binds to $3'$) through the coordinated force of multiple hydrogen bonds at each base-pairing, where A binds to T and C binds to G. [10]

DNA is able to undergo duplication, which occurs through the coordinated action of many molecules, including *DNA polymerases* (synthesizing new DNA), *DNA gyrases* (unwinding the molecule), and *DNA ligases* (concatenating segments together).

Transcription of DNA into RNA   In order for the genome to direct or effect changes in the cytoplasm of the cell, a transcriptional program may be activated for the purpose of generating new proteins to populate the cytosol—the heterogenous intracellular soup of the cytoplasm. DNA remains in the nucleus of the cell, while most proteins are needed in the cytoplasm of the cell, where many of the cell's functions are performed. Thus, DNA is copied into a more transient molecule

---

[10]These pairings are present in all DNA and are the most thermodynamically stable of all possible pairings of nucleotides, which accounts for the high specificity with which complementary strands of nucleotide polymers bind to each other.

called RNA. [11] A *gene* is a single segment of the coding region that is transcribed into RNA. RNA is generated from the DNA template in the nucleus of the cell through a process called *transcription*.[12]

The RNA sequence of base pairs generated in transcription corresponds to that in the DNA molecules using the complementary A-T, C-G, with the principal distinction being that the nucleotide uracil is substituted for the thymine nucleotide. Thus, the RNA alphabet is *ACUG* instead of the DNA alphabet *ACTG*. Each cell contains around 20 to 30 pg of RNA, which represents 1% of the cell mass. The RNA that codes for proteins is called *messenger RNA*, and the part of the DNA that provides that code is called an *open reading frame* (ORF). When read in the standard 5′ to 3′ direction, the portion of DNA before the ORF is considered *upstream*, and the portion following the ORF is considered *downstream*.

The specific determination of which genes to transcribe is determined by *promoter regions*, which are DNA sequences upstream of an ORF. Many proteins have been found containing parts that bind to these specific promoter regions, and thus activate or deactivate transcription of the downstream ORF. These proteins are called *transcription factors*.[13]

A diagram of the genetic information flow, from DNA to RNA to protein, is illustrated in Figure 1.7

Prokaryotic and eukaryotic cell types   Although there are many taxonomies one could use, we can essentially divide the world of organisms into two types: *eukaryotes*, whose cells contain compartments or organelles within the cell, such as mitochondria and a nucleus; and *prokaryotes*, whose simpler cells do not have these organelles. Animals and plants are examples of eukaryotes, while bacteria are prokaryotes. Most prokaryotes have a smaller genome, typically contained in a single circular DNA molecule. Additional genetic information may be contained in smaller satellite pieces of DNA, called plasmids.

---

[11]That portion of the entire DNA molecules that is transcribed into RNA is called the *coding region*.

[12]Transcription involves unwinding a DNA molecule so that the particular gene that is to be transcribed is sufficiently exposed to the transcriptional machinery, notably RNA polymerase.

[13]Not all RNA codes for proteins, however. In fact, only 4% of total RNA is made of coding RNA. Of the noncoding RNA, ribosomal RNA (rRNA) and transfer RNA (tRNA) are used in various components of the protein translational apparatus mentioned below, and are not themselves translated into proteins. Eukaryotes also contain small nuclear RNA (snRNA), which is part of the splicing apparatus (see below); small nucleolar RNA (snoRNA), which is involved in methylation of rRNA; and small cytoplasmic RNA (scRNA), which can play a role in the expression of specific genes.

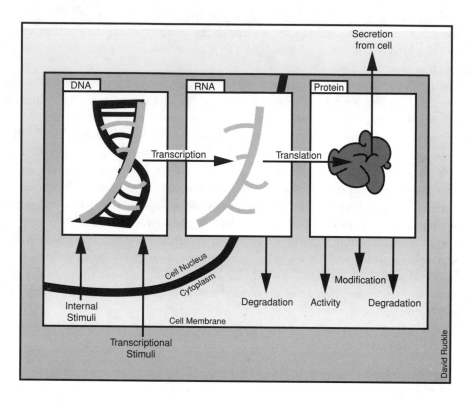

Figure 1.7
Flow of genetic information, from DNA to RNA to protein. This simplified diagram shows how
the production of specific proteins is governed by the DNA sequence through the production of
RNA. Many stimuli can activate the specific transcription of genes, and proteins can play a wide
variety of roles within or outside cells. Note that even in this simplified model, it is obvious that
since we are currently able to (nearly) comprehensively measure only gene expression levels, we
are missing comprehensive measurements of protein modification, activity, transcriptional
stimuli, and many other components of the state of a cell.

The structure and processing of RNA transcripts Eukaryotic genes are not necessarily continuous; instead, most genes contain *exons* (portions that will be placed into the mRNA) and *introns* (interruptions that will be spliced out). Functions have been recently discovered for introns, such as promoter-like control of the transcription process. Introns are not always spliced consistently; if an intron is left in the mRNA, an *alternative splicing product* is created. Various tissue types can flexibly alter their gene products through alternative splicing. [14]

Before coding RNA is ready as mRNA, the pre-mRNA must be processed. [15] In eukaryotes, after the splicing process, the generated mRNA molecule is actively exported through nuclear pore complexes into the cytoplasm. The cytoplasm is where the cellular machinery, in particular the ribosomal complex,[16] acts to generate the protein on the basis of the mRNA code. A protein is built as a polymer or chain of amino acids, and the sequence of amino acids in a protein is determined by the mRNA template. The mRNA provides a degenerate coding in that it uses three nucleotides to code for each of the twenty common naturally occurring amino acids that are joined together to form the polypeptide or protein molecule. With three nucleotides there can be 4^3 possible combinations for a total of 64 combinations. Consequently, several trinucleotide sequences (also known as codons) correspond to a single amino acid. There is no nucleotide between codons, and a few codons represent start (or initiation) and stop (or termination).[17] The process of generating a protein or polypeptide from an mRNA molecule is known as *translation*.

As an example, if an RNA transcript had the nucleotide sequence

GCT TGC AAG GCG

(the spacing and grouping by three, which obviously is not visible in the RNA transcript, was added to make this easier to read), the first three nucleotides would code for this chain of amino acids

Alanine Cysteine Arginine Alanine

Note that both the initial GCT and the final GCG code for alanine; because of the degenerate coding mentioned above, there are four codes for alanine in the standard genetic code.

---

[14] In fact, some cells can use the ratio of one alternative splicing to another to govern cellular behavior.

[15] Modifications to the RNA include the addition of a cap at the $5'$ end and a tail made of repeated adenine nucleotides (the poly-A tail).

[16] A complex containing hundreds of proteins and special-function RNA molecules.

[17] There are notable exceptions: the code for the naturally occurring amino acid selenocysteine is identical to that for a stop codon, except for a particular nucleotide sequence further downstream.

Most mRNA has a terminating poly-A tail. This terminal sequence makes it easy to pick out the labeling reactions that are used before hybridization to DNA microarrays, described in section 3.1.

Processing of amino acid chains   Once the protein is formed, it has to find the right place to perform its function, whether as a structural protein in the cytoskeleton, or as a cell membrane receptor, or as a hormone that is to be secreted by the cell. There is a complex cellular apparatus that determines this translocation process. One of the determinants of the location and handling of a polypeptide is a portion of the polypeptide called the *signal peptide*. This header of amino acids is recognized by the translocation machinery and directs the ribosomal-mRNA complex to continue translation in a specific subcellular location, *e.g.*, constructing and inserting a protein into the endoplasmic reticulum for further processing and secretion by the cell. Alternatively, particular proteins may be delivered after translation and chaperones can prevent proper folding until the protein reaches its correct destination.

Transcriptional programs   Initiation of the transcription process can be caused by external events or by a programmed event within the cell. For instance, the piezoelectric forces generated in bones through walking can gradually stimulate osteoblastic and osteoclastic transcriptional activity to cause bone remodeling. Similarly, heat shock or stress to the cell can cause rapid change or initiation of the transcriptional program. Additionally, changes in the microenvironment around the cell, such as the appearance of new micro- or macronutrients or the disappearance of these, will cause changes in the transcriptional program. Hormones secreted from distant organs bind to receptors which then directly or indirectly trigger a change in the transcriptional process.

There are also fully autonomous, internally programmed sequences of transcriptional expression. A classic example of this is the internal pacemaker that has been found with the *clock* and *per* genes (see figure 1.8 below) where, in the absence of any external stimuli, there is a recurring periodic pattern of transcriptional activity. Although this rhythmic pattern of transcription can be altered by external stimuli, nonetheless it will continue initiating this pattern of transcription without any additional stimuli.

Finally, there are pathological internal derangements of the cell which can lead to transcriptional activity. Self-repair or damage-detection programs may be internal to the cell, and can trigger self-destruction (called apoptosis) under certain conditions, such as irreparable DNA damage. As another example, there may be a deletion mutation of a repressor gene causing the gene normally repressed to instead be highly active. There are many clinical instances of such disorders, such as familial male precocious puberty [161] where puberty starts at infancy due to a mutation in the luteinizing hormone receptor. This receptor normally activates only when luteinizing hormone is bound, but with the mutation present, activation does not require binding.

1.5.1 Biological *caveats* in mRNA measurements

There are several desiderata that should be understood about transcription in order to better understand the limitations of the gene profiling techniques.

- *The set of protein-coding RNA, otherwise known as the* transcriptome, *should be viewed as a pool of mRNA.* Even at equilibrium, each type of mRNA is degraded at its specific rate. New transcription provides additional mRNA to replace those being degraded, maintaining the levels. Thus, a transcriptional program should more accurately be viewed as a change in the transcriptome. In addition, just because a particular mRNA is seen in the transcriptome, it does not necessarily imply that the mRNA is about to be translated. Similarly, presence of a particular mRNA in the transcriptome does not mean that mRNA was recently transcribed. To determine what genes may be in the process of being translated, one can specifically look at polyribosomal mRNA, or that mRNA found in the presence of ribosomes. To determine which genes are currently being transcribed and processed, one can construct studies looking for pre-mRNA levels instead of mRNA levels (*i.e.*, by finding intronic splice sites, *etc.*).

- *mRNA can be transcribed at up to several hundred nucleotides per minute but may be transcribed at much slower rates.* In eukaryotic organisms, genes can take many hours to transcribe. For instance, a gene found in muscle, dystrophin, can take up to 20 hours to be transcribed. The lifespan of an mRNA molecule from initiation of transcription to ultimate degradation is a complex function of the rate of transcription, the stability of the particular mRNA molecule, and changes in its processing due to other cellular events, whether internally or externally initiated. Gene expression measurements, as performed by microarrays, measure the concentration of each mRNA as a snapshot at a single point of time or relative to another sample of mRNA. If the transcription of two genes is simultaneously stimulated and proceeds at the same rate, but the mRNA samples have different lifespans, then the expression measurements (and downstream protein production) from these two genes can be quite divergent.

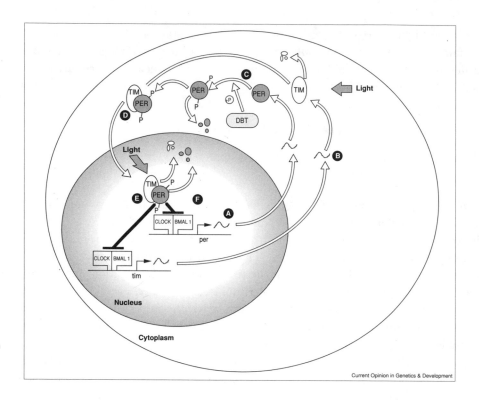

Figure 1.8
Genetic machinery of the circadian rhythm. Current molecular model of rhythm generation in
Drosophila, from [191]. The succession of events (A–F) occur over the course of approximately
24 hours. *A*, CLOCKBMAL heterodimers bind the *per* and *tim* promoters and activate mRNA
expression from each locus; CLOCKBMAL may also activate transcription of other
circadian-regulated genes (not shown). *B*, *per* and *tim* mRNA are transported to the cytoplasm
and translated into PER and TIM protein, respectively. *C*, Regulation of protein levels occurs
by two mechanisms: DBT protein phosphorylates and destabilizes PER, and light destroys TIM.
Light during the early subjective night can phase-delay the clock. Small "blobs" indicate
degraded proteins. *D*, PER and TIM levels slowly accumulate during the early subjective night;
TIM stabilizes PER and promotes nuclear transport. *E*, PER and TIM dimers enter the nucleus
and inhibit CLOCKBMAL-activated transcription. *F*, Protein turnover (combined with the lack
of new PER and TIM synthesis) leads to derepression of *per* and *tim* mRNA expression; the
cycle begins again (*A*). Light during the late subjective night can phase-advance the clock.

- *Proteins perform most cellular functions.* The lifespan of proteins is at least as variable as that of mRNA. Consequently, measurements of gene expression (*i.e.*, mRNA measurement) may not accurately correspond to the concentration or activity of the protein for which it codes. This should be a caution to any investigator imputing function to a gene based solely on gene expression patterns.

- *The mRNA generated from the transcription of a gene may differ depending on which exons and introns are or are not spliced into the mRNA molecule.* These alternate splicing products are mRNA molecules with divergent sequences. Therefore, measurements of mRNA that are designed to measure only one of these alternate splicing products will provide incomplete if not misleading information.

- *The genetic basis for organismal diversity is due in large part to differences in sequences, also known as polymorphisms, of each gene.* Most of these polymorphisms differ from one another by one nucleotide and are known as single nucleotide polymorphisms (SNPs). Due to the small portion of the genome coding for proteins and to redundancy in the mRNA code, described on page 27, only some SNPs will result in differently constructed proteins. If a gene expression measurement technology is highly specific for a particular SNP, then other variants will not be measured. When we consider that the Human Genome Project to date only includes the sequence of handfuls of individuals, the implications of such measurement specificity become apparent. Nonetheless, as common SNPs become documented for each gene, it is likely that successful expression measurement techniques will measure each of them.

Despite these *caveats*, it remains that gene profiling studies have proven to be remarkably robust in describing the functional grouping and coordinated behavior of genes. The reasons for this are discussed in section 2.2, page 60.

1.5.2   Sequence-level genomics

Sequencing or genotyping the genome of individuals allows the characterization of what distinguishes the hereditable material of each individual from that of others. By matching differences in phenotype (*e.g.*, blood pressure, adult height) between individuals and this genomic characterization (*i.e.*, in association studies) genetic epidemiologists are able to impute these phenotypic differences to a small span of the genome. The smallness of the span is a function of the spatial resolution with which the genomic characterization occurs. Although there is controversy around what constitutes sufficient resolution, there is some consensus that genomic markers such as SNPs spaced every thousand bases will be sufficient to unambiguously resolve the span of the genome associated with a phenotypic difference to a single gene [112]. Currently, the cost of a single genotype is around \$0.50 and so the cost of a high-resolution genome scan of an individual is on the order of magnitude of \$1 million. At this cost, only a very few institutions can afford a comprehensive study of a population. If the recent past is to be a guide, the cost of genotyping is likely to drop by several orders of magnitude well within a decade, at which point genome-wide scans of populations will become economically feasible.

The kind of information that these studies will provide includes the contribution of particular polymorphisms to changes in phenotype, presumably via changes in gene function, pattern of expression, or both. Currently, expression microarrays capture information about gene polymorphism poorly, if at all. Presently shipping microarrays typically code for a canonical "normal" gene sequence. Departures from this sequence will result in changes in the intensity reading reported by any of the currently employed microarray platforms, as will become obvious after reading the chapter on measurement techniques (chapter 3). Undoubtedly, in the next 5 years, the continued geometric increase in the complexity and density of expression microarrays will allow specific assaying for all common and all clinically significant polymorphisms. Until then, however, polymorphisms will be a source of "noise" or unexplained variation. Functional genomics investigations designed to elucidate the role of particular polymorphisms in phenotypic mechanisms or phenotypic variation are best conducted using methods other than RNA expression detection microarrays, such as high-throughput sequencing and genotyping. These latter techniques are beyond the scope of this text (see [43, 112] for an excellent survey of sequencing and genotyping) and therefore this particular avenue of functional genomics will not be addressed below except tangentially.

1.5.3 Proteomics

There is a good deal of enthusiasm at present about the emerging discipline of proteomics. The promise of proteomics is that we will be able to measure, in a similarly comprehensive and parallel fashion to RNA microarray measurements, the concentrations of proteins present in a particular cellular system. In a very abstract sense, for the purely computational bioinformatician, the field of proteomics does not harbor any particular novelty in that all it provides is another 100,000 variables, or more, to describe the state of cellular processes. In that perspective, a proteomics data set reduces simply to another array that has distinguishing noise characteristics (*i.e.*, sources of biological and measurement variability as described in section 3.2.5) and is just as amenable to the clustering and classification techniques of chapter 4 as any set of microarray expression measurements. Less abstractly, however, proteomics offers a set of insights that are quite different and divergent from those of expression microarrays. The assumption underlying expression microarray measurements is that by capturing the patterns of expression management, we will capture the basic irregulatory rhythms of the cell [34]. Although these assumptions may hold at times and have done remarkably well in helping biologists elucidate some fundamental biology and to classify clinical phenomena, there are several persuasive reasons why these assumptions should not always hold. First, we know that most of the effector molecules in cellular metabolism are proteins. To the extent that the timing of protein synthesis and the half-life of proteins is not closely coupled to that of RNA expression, the assumption of the representativeness of RNA levels does not hold. As outlined in section 1.5, these assumptions do not hold in many instances. Nonetheless, in proteomics, we will be bedeviled by a new set of assumptions that will be equally problematic and challenging. Assuming that we have high reproducibility and compact systems for assessing the concentration of tens of thousands of proteins, we will be faced with the following challenges:

- Similar concentrations do not imply co-regulation. Given that proteins have hugely different half-lives, even within a single cell (*e.g.*, a structural protein in a bone osteoblast and a parathyroid hormone receptor in the same cell), then the concentrations of protein molecules in a cell may only remotely reflect joint regulation. This problem also haunts the analysis of RNA expression microarray data because of the wide range of the stability-degradation rate of mRNA.

- Conversely, repeatedly different concentrations of two proteins imply co-regulation. At any given sampling time, the two proteins could have quite variable concentrations and different mutual relationships. Yet, there is nothing about this to preclude important functional interactions between these proteins.[18]

- Localization heterogeneity. Unlike transcription of genes, which occurs within the nucleus, protein activity has very distinct and heterogeneous functional significance in different parts of the cellular compartments, and therefore an essential part of understanding protein function and regulation from proteomic data will require detailed localization to subcompartments of organelles in order to be meaningful. This problem also exists with RNA expression microarray data, because of the differing biological implications of RNA concentrations measured before splicing, after splicing in the cytoplasm, and during translation.

---

[18]One way to circumvent this problem is with multiple serial measurements of all proteins and then to subject them to a dynamics analysis.

These challenges of proteomics will eventually be addressed by novel ways of looking at protein activity over time and in different spatial locations. Nonetheless, at present, the basic mechanism for cheaply and reliably obtaining large numbers of parallel measurements of protein activity have yet to be worked out and industrialized, so that these developments are not likely to occur on a large scale for at least 1 or 2 years. When these challenges have been resolved, then indeed the arrays of proteinomic data will be amenable to the same techniques of analysis as described in this book for RNA expression. For thoughtful and comprehensive insights into the challenges of proteomics and data analytic techniques, we refer the reader to the following papers [25, 69, 141] and websites `http://www.expasy.ch/`, `http://www.hip.harvard.edu/`.

2 Experimental Design

2.1 The Safe Conception of a Functional Genomic Experiment

Experimental biologists and biomedical scientists who are considering the use of microarrays in their research programs have often asked us questions of the following form: "How do I best design an experiment with the following mutant mouse?", or, "We have a budget for 20 microarrays to analyze this system. How can we complete the experiment within budget?" Unlike the typical biomedical study design, many of these investigations that the experimentalist are contemplating are driven not by the testing of a particular hypothesis, but by the hope of generating hypotheses that will then lead to a directed and relevant experiment that will subsequently yield an interesting biological or biomedical insight.

We find it even more challenging when we are asked secondary questions related to the subsequent microarray data analyses such as, "What is the right machine-learning technique or clustering technique to apply to a set of data to best obtain an answer about [some biological mechanism]?" The questioner often goes on to ask whether one particular software package works better than another, is more reliable, or whether it is worth its cost. We view these questions as well-intentioned but fundamentally flawed and ill-posed. Although a particular machine-learning algorithm may be slightly superior to another in one set of circumstances (see section 2.2.2 on the relative merits of different clustering algorithms), the key to a successful functional genomic investigation is to start with an experimental design that maximizes the possibility of observing gene expression patterns that are relevant to and informative of the biological aspect or question under investigation. If the experimental design is such, then *most* of the common machine-learning algorithms will reveal those patterns.

So how does one proceed to an effective experimental design for the expression of thousands of genes in order to, hopefully, answer a biological question? We have found it useful to frame our answers within the concepts of *experiment design space* and *expression space*. Let us start with the experiment design space.

2.1.1 Experiment design space

Experiment design space is the space of potential combinations of stimuli or conditions to which one can subject the regulatory apparatus of the genome. These stimuli include exposure to pharmaceutical agents such as mitogens, environmental effects such as temperature differences, withdrawal of nutritional sources, overexpression of a particular gene such as a transcription factor in a transgenic animal, or conversely, an engineered misexpression or "knockout" model. Other common experimental conditions include the passage of time, such as from a newborn to an old mouse; the observation of the environment (organisms from one particular condition vs. another); the observation of organismal processes (organisms with a particular morphology, or afflicted with a particular pathophysiologic condition vs. another); physical manipulation of the cell membrane; exposure to infectious agents; and changes in osmotic conditions, all of which can potentially affect the genomic regulatory mechanisms which maintain cellular existence and the regulation of the cellular response to the environment.

Experimental conditions can be represented as combinations of values on several scales:

- *Binary* scales, *e.g.*, the presence or absence of a diagnosis.

- *Continuous* scales, *e.g.*, such as the gradients of exposure to a chemical. These can have value ranges, say, 10.0, 1.0, 0.1, 0.01, representing the chemical concentration or 0, 5, 10, 15, 20, representing time in minutes from the initial state.

- Unordered *discrete* scales, *e.g.*, the type of mutant mouse model created or which of the multiple mammalian muscle tissue types were sampled.

When different experimental conditions from these different scales are combined, then the experiment can be described as being located in a particular locus of the multidimensional experiment design space. Shown in figure 2.1 below is the simple example of an experimental space over three sets of conditions. In this case, the first dimension represents whether the mouse model being studied is an *irs1* [138] knockout, an *irs2* knockout [192], or a wild-type mouse. The insulin receptor signaling proteins *irs1* and *irs2* are thought to be part of the signaling cascade through which multiple cellular events[1] are triggered upon binding of insulin to the cellular receptor. The "knockout" is an engineered mouse in which a particular gene is no longer sufficiently expressed. The second dimension is the amount of insulin to which the mouse circulatory system is exposed. The third dimension represents the three types of mouse tissues being studied in this particular set of experiments, namely, the liver, brain, and fatty tissue. Each of these tissues has distinct metabolic responses to insulin exposure. Within this three-dimensional experiment design space, each set of possible experiments can be uniquely located. This notion obviously generalizes to the much larger number of possible dimensions in any experimental system. Note that the first and third dimensions have unordered discrete scales, whereas the second dimension is continuously scaled.

---

[1]For example, cellular proliferation and differentiation during development.

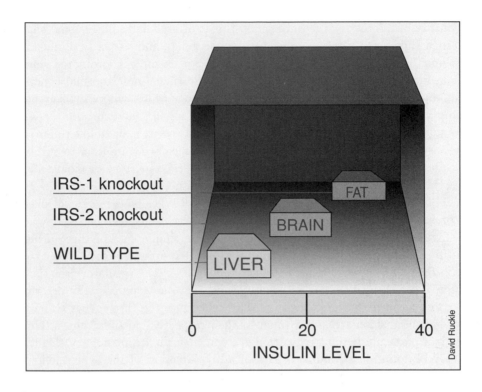

Figure 2.1
Three experiments within the multi-dimensional representation of experiment design space.
Experiment design space defines all the possible stimuli or conditions to which a particular
biological system could be subjected. Shown here is an experiment design space that is
concerned with insulin signaling in different tissues, in different mouse "knockout" models, with
different levels of insulin.

A functional genomic study might involve a trajectory across the experiment
design space where the mouse models are studied across multiple insulin levels and
multiple tissues. As the number of experimental dimensions increases, the number
of experiments required to cover the experimental design space grows exponentially.
However, the concept of an experiment design space by itself does not suffice to
determine which kind or number of experiments is adequate to assess the genomic
physiology of a tissue or system.

2.1.2 Expression space

Now that we have defined the experiment design space, we define the expression
space to be the space of potential expression values of all genes in a given genome. In

the context of microarray measurements, this is a large multidimensional space—with each dimension corresponding to a single gene—and it can be immediately partitioned into several overlapping subspaces. For example, coordinated gene expression measurements made from normal muscle tissue under normal conditions can be represented by a subspace of expression space which will most likely have a nonempty intersection with the subspace of coordinated gene expression measurements made from diseased muscle tissue. We often repeat mantra-like to ourselves the phrase "A cell is a cell is a cell," by which we mean that in order to survive as living and organized entities, most cells share a large repertory of biological processes, whether they be components of muscle tissue, liver tissue, or white blood cells. That is, the expression subspaces of most cell types have significant overlap.

Expression space contains clues about the interactions between two or more clusters of genes. As an example, we start with the simple case of three arbitrary genes: insulin-like growth factor I (*IGF-1*), insulin-like growth factor binding factor-1 (*IGF-BP1*), and fibroblast growth factor 2 (*FGF2*). The regulatory relationship between these three genes[2] is, in some tissues, that one gene partially determines the expression level of the other two. This will cause the three genes to move in a coordinated fashion across the range of their respective possible values. That is, if this is a particularly tight regulation system, then the expressions of this triplet will only be found in a small subspace of all possible loci of the expression space. If this were a loosely regulated or unregulated system, it is possible that the gene expression values might be found in all parts of the expression space, as shown in figure 2.2 below. The totality of the potential loci that these genes can occupy constitutes the potential set of all coordinated gene expression levels, or expression space. The subset of the expression space wherein a cellular system has its expression under all stimuli is called the *transcriptome* [41, 184]. More precisely, the transcriptome encompasses the expression space of all putative 30,000 genes in the human genome (or 100,000 gene products if current alternative splicing estimates are correct). Each point in the human transcriptome corresponds to the possible values that each gene's expression might attain under any physiological or pharmacological stimulus.

The meta-goal of the experimentalist/investigator then is to define the experiment design space that provides the best and most complete picture of the entire transcriptome *or* a subset of the expression space of interest. This is the point in experimental design where a great deal of domain-dependent knowledge, *i.e.*, specific or *a priori* biological knowledge, is of significant utility. The investigator must

---

[2] All of these genes have been implicated in somatic growth and bone repair.

ask herself or himself: "What is the minimal set of experimental conditions, *i.e.*, the smallest subspace of the *experiment design space*, that will cause the largest set of relevant subspaces of the *expression space* to be sampled?" In order to answer this question, the investigator must first understand how each experimental condition could potentially affect the expression of potentially interesting genes and be able to quantify its effect. Second, the investigator must consider all possible subspaces or interactions between genes that are of interest. For example, let us say that we are only interested in the interactions between *IGF-1* and *IGF-BP1* in the example earlier in the section and we wish to determine how the relationship between these two genes changes under both physiological and pharmacological conditions. Then we would obtain those sets of experiments in the experiment design space which would most likely cause one or both of these genes to be expressed over the broadest range of their possible values so that all the evidence of the interactions would be captured, *i.e.*, the subspace of experiments that covers the largest plane in the *IGF-1* and *IGF-BP1* expression subspace. A researcher might be interested in the signaling mechanisms of *IGF-1* once it has bound to *IBF-BP1* and how it affects the liver's expression of other genes. These genes may code for many proteins involved in postreceptor signaling, including possibly *IGF-BP1* itself. With a finite budget and a finite source of tissue or animals, the investigator will have to choose between alternative strategies. A "mutant screening" strategy might be to assay expression levels in the liver over a set of knockout mice in which each element of the set constitutes a knockout of a different putative element of the *IGF-1* signaling pathway, whereas a "physiological stimulus" strategy, would be to take a single element of the set (one of the knockout strains or a wild-type animal) and subject it to a sequence of different concentrations of extracellular *IGF-1*. Determining which of these alternatives provides the best exploration of the expression space will require an understanding of whether the range of mutants or physiological stimuli engages the portion of the genetic machinery that is relevant, and how close to the normal homeostatic behavior or pathological behavior of interest is the experiment intended to explore.

Figures 2.3, 2.4, and 2.5 illustrate the pitfalls of insufficiently exercising the regulatory mechanisms of gene expression. In figure 2.3, we observe the joint behavior of two genes in which there exists a strong component of randomness in the relationship between the two genes—as evident from the low Pearson's correlation coefficient that is calculated between these two genes within the current range of expression levels. This is also summarized by the Pearson's correlation coefficient for the linear regression calculated between these two genes, which is calculated to be very low. However, if an additional stimulus is applied to the regulatory system

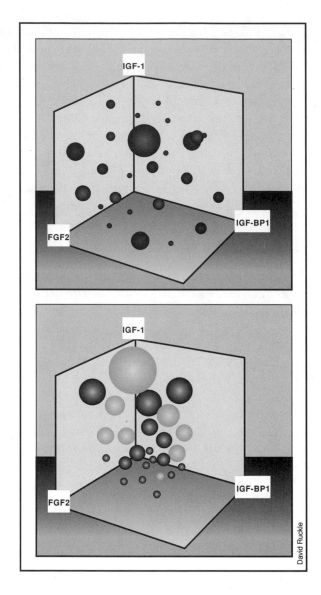

Figure 2.2
Expression space in which gene expression is loosely versus tightly coupled. If genes are tightly coupled in the expression space, they will tend to occupy a small subspace of expression space over any set of experiments. If, however, these genes are loosely coupled or even causally unrelated, then their expression levels will have little relation to one another and therefore tend to be scattered over a greater volume of the expression space.

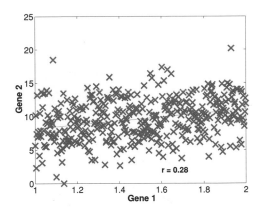

Figure 2.3
Apparently random relationship in expression space between two genes.

such that gene 1 has a wider range of expression, then despite the randomness in the relationship between the two genes observed in the previous range of expressions, we begin to see what appears to be a linear relationship with a much larger Pearson's correlation coefficient (figure 2.4). If the expression level of gene 1 is perturbed to an even wider range, it becomes apparent that the relationship is not linear but, in fact, curvilinear, possibly even a quadratic relationship, as shown in figure 2.5. If the experiment space was incorrectly selected, then these and many other pairs of genes like it would be assumed to be poorly or randomly related when, in fact, with the right stimulus, a tight curvilinear relationship would be determined, and *vice versa*.

2.1.3   Exercising the expression space

In section 2.1.1, we described how each experiment can be viewed as an exploration of the space of all possible expression patterns. In exploring this expression space, the investigator is constrained to the space of possible experiments. Informally, the goal of experiments should be to *maximally exercise the genome* bringing out the couplings or correlations between genes that are of greatest relevance to the biological process being studied. We have found the following "watch metaphor" to be of some use in explaining the interaction between expression space and experiment design space. We leave it to the reader to judge whether it is helpful, or simply a confusing distraction.

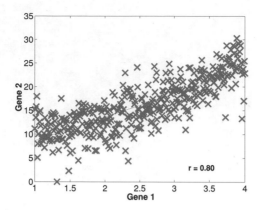

Figure 2.4
Apparently linear relationship in expression space between two genes.

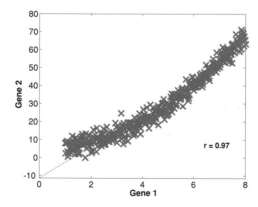

Figure 2.5
Apparently curvilinear relationship in expression space between two genes. This curvilinear relationship would be obscured if the expression space had been insufficiently exercised.

The watch metaphor   Suppose, that we were interested in discovering how an old analog watch of the sort illustrated in figure 2.6 worked. Simply watching the hands of the watch going around might allow us to generate some hypotheses about the mechanisms that generate this motion or behavior. Empirically, we could derive the relationship between the movement of the minute hand and the hour hand, but the total sum of mechanisms that could explain the behavior would be large. A more invasive investigation would be to take a hammer to the watch and examine the pieces that fall out of the watch's case (figure 2.7). The combination of the prior observed empirical behavior and the examination of these pieces might give us a better, if still grossly incomplete hypothesis (not to mention the loss of a costly watch). That is, the number of possible mechanisms explaining the observed behavior of the watch hands would have been substantially reduced. Furthermore, if one had no other similar watch, these hypotheses might not be verifiable on account of its destruction. A more sophisticated approach might be to pry open the back of the watch and observe the movement of the gears and escape mechanisms. With these observations, and one or both of the prior kinds of investigations, an even better understanding can be obtained of how the various components of the watch are mechanistically related to one another. All of these experiments are observational studies, and at best we could generate some fairly accurate hypotheses without ever demonstrating causality. Causality would be most convincingly demonstrated by intervening in the operating mechanism, *e.g.*, winding the watch (partly or until the mechanism breaks), shaking the watch, immobilizing one of the gears, or replacing a gear with one of different-sized teeth. These interventions are progressively more technically challenging but also provide better insights and often confirmation of the hypotheses generated during the earlier observational studies.

The analogy with expression studies, with the indulgence of the reader, can be made as summarized in table 2.1. These describe different kinds of experiments—distinct points in the potential experiment space. Some of these experiments allow the components of the watch to work in their usual fashion, *i.e.*, within their usual "expression space," whereas others bring them into an operating mode or expression space in which they do not usually function. The question the watch investigator would have to ask himself or herself is what part of the expression space of the watch will best inform him or her of the normal function of the watch. It may be that smashing the watch is the only part of the experiment space that is cost effective or feasible for the investigator. The smashed or overwound state of the watch may not accurately reflect the normal intact and possibly more interesting state of the watch's expression space.

To take these analogies one step further, there are multiple arrangements of gears

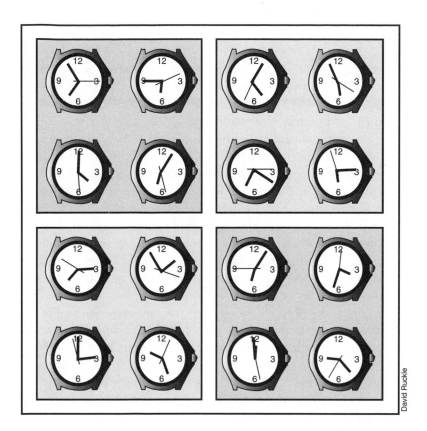

David Ruckle

Figure 2.6
Noninvasive monitoring of mechanistic behavior. By observing the concerted action of the watch hands, several competing *weak* hypotheses about the underlying mechanism can be generated. The more observations made at different times, the smaller the number of possible hypotheses generated.

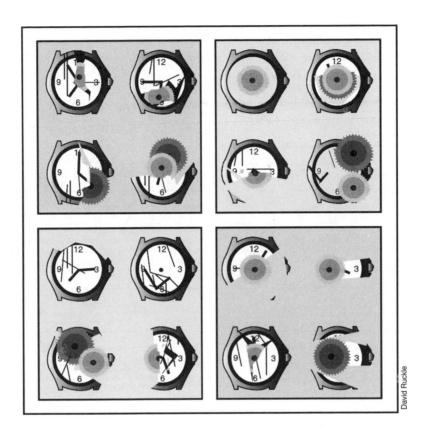

Figure 2.7
Decomposition of the watch. An invasive exploration of mechanism will reveal most of the components of that mechanism. In the process of the invasive investigation some of the components will be damaged. Also, the relationship of these components inside a working watch must be inferred and is not directly observed because of the invasive nature of this investigation.

Experiment Space in Watch	Experiment Space Biological Analogy
Observing the movement of the hour, minute hands	Hypothesizing cellular function based on gross cellular behavior
Smashing the watch, examining the pieces	Sequencing some of the genes in the cell or measuring a few proteins or RNA transcripts
Seeing all the gears working from the back of the opened watch	Parallel measurement of thousands of genes at a time Time series measurements
Partially winding up the watch	Physiological stimulus
Winding the watch until it is near breaking or until it breaks, thereby bringing the internal gears of the watch into behavior they would not typically exhibit	Pharmacological stimulus that brings the gene expression patterns into an operating region outside the usual physiological range
Removing a gear	Creating an organism with a gene "knocked out"
Replacing a gear with one of different weight	Creating a transgenic organism

Table 2.1
Exploring the experiment space with a watch

that could account for the observed interactions between these gears and their effect on the hour hand and minute hand. Similarly, although perturbations of the genome can provide insight into the interactions among the major genes involved in a given pathological pathway, they do not reveal the entire pathway. Rather they expose a significantly large family of possible mechanisms that could account for the observed behavior. We address this issue in greater depth in the section on pathway re-engineering (section 4.13.2). Nevertheless, as suggested earlier, there are some misconceptions about what can be done on the basis of microarray expression studies implicit in some of the current genomic and bioinformatics literature.

Misconceptions in "reverse engineering"   To date, there have been several efforts to try to reconstruct pathways from gene expression measurements [57]. Given the success of recapitulating the first few steps of the glycolytic pathway from substrate measurements [12], it would appear plausible to do this. However, without careful consideration of the pathways that one is trying to reconstruct, there is significant risk of a methodological and metaphorical flaw in this analysis.

It is important to realize that pathways could exist on several molecular and physiological levels. The genes that are regulated in one pathway may play a role in another pathway. As an example, let us consider the genes coding for lactic acid dehydrogenase: The gene *LDHA* is expressed mostly in muscle and is located on chromosome 11p15.4, while *LDHB* is located on chromosome 12p12.2 and is expressed mostly in heart. *LDHA* is known to have binding sites for HIF-1, CREB-1/ATF-1, and p300/CREB binding protein [62], as well as binding sites for NRE, NREBP/c-FOS, AP1, AP2, and other transcription factors in its promoter region [104]. *LDHB* is thought to have a binding site for SP1 [159]. There is also likely to be a post-transcriptional level regulation of these two genes. Because of the differences in promoter regions, it is safe to assume that both *LDHA* and *LDHB* participate in their own gene expression regulatory networks.

Both *LDHA* and *LDHB* code for protein subunits. The final protein product is made up of four subunits. Various combinations of the two types of subunits are assembled into five different lactic acid dehydrogenases, from LDH-1 (containing four *LDHB* subunits), to LDH-5 (containing four *LDHA* subunits). These five enzymes are found in a binomial distribution in mammals. Additionally, the final protein itself is an enzyme that assists in the conversion of lactic acid to pyruvate, called lactic acid dehydrogenase, and defined by the Nomenclature Committee of the International Union of Biochemistry and Molecular Biology as Enzyme Commission (EC) 1.1.1.27. Lactic acid and pyruvate happen to be substrates in the glycolytic pathway of anaerobic energy production.

Thus, in this one simple example, one finds at least three pathways, as shown in figure 2.8. The operator at the *enzyme level pathway* participates as one enzyme in a major biochemical substrate pathway, that of glycolysis. That operator, however, is assembled stochastically in the *assembly pathway* from subunit components. Finally, those subunits are produced within their own *genetic regulatory pathways* involving a multitude of transcription factors. Considering the three layers of pathways, it should now become apparent that it may not be possible to "reconstruct" the glycolytic pathway simply from measuring multiple gene expressions. However, on a more optimistic note, given enough gene expression data, one should be able to reconstruct the first layer of these pathways, *i.e.*, that of gene expression regulation and involving transcription factors and promoter regions.

How vigorously should a model system be stimulated or perturbed? Let's revisit the phrase "maximally exercising the genome." What constitutes a maximum perturbation will in part be defined by the goals of the investigator. If she or he wishes to understand the interactions between the genes under normal physiological conditions, then it may be that only relatively minor perturbations followed across time will be informative. Even though extreme perturbations are easier to measure, they might not accurately represent interactions that occur over the normal dynamic range of activities of the RNA species, structural proteins, and enzymes programmatically generated by the genome. There are, however, several circumstances when it may be advantageous to perturb the relationships between the various constituencies of the transcriptome much more violently and dramatically than is found under normal physiological or homeostatic conditions. An example of such a motivation might be to measure the effect of pharmacological doses, rather then physiological doses, of a pharmaceutical. The intended effect may lie outside the normal operating range of the genetic machinery, as is often the case in pharmacology. Other times, the goal may be to examine biological systems which naturally fall far outside the usual operating parameters of the genomic machinery. This includes a variety of disease states such as a malignancy or diabetes, where there are single or multiple massive perturbations of the usual homeostatic system, and therefore a perturbation of the model system that corresponds to the pathological condition would be appropriate. It is worthwhile considering some extreme examples to illustrate the importance of this decision.

For example, if one is trying to understand the relationship between a set of genes thought to be involved in the effect of insulin signaling, it may not be particularly helpful to start with a sample population of human adipose tissue (fat) which is known to be insulin sensitive, but obtained from individuals who are all

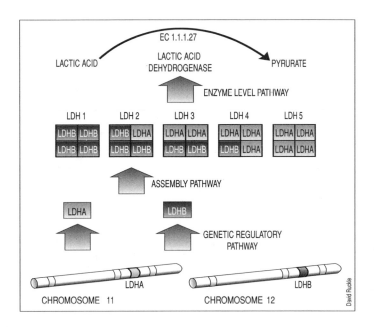

Figure 2.8
An example of why discovery of pathways solely using gene expression measurements is difficult. At least three "pathways" are involved in the conversion of lactic acid to pyruvate. The highest enzyme level pathway involves the action of an enzyme called lactic acid dehydrogenase, designated EC 1.1.1.27. However, when one traverses to a lower-level pathway, one learns that the role of this enzyme is performed by five separate protein products, which appear in a binomial distribution. When one traverses lower, through the assembly pathway, one learns this distribution is present because of the various possible combinations of *LDHA* and *LDHB*, two individual protein subunits, of which four must be put together to make an active enzyme. Then, when one traverses even lower, through the genetic regulatory pathway, one learns that each of these subunits is regulated differently, and appears on different chromosomes. Not shown are the pathways involved in transcription, translation, movement of proteins to the appropriate locations, and movement of substrates into and out of the cells and organelles, and many other pathways.

in the fasting state when insulin levels are low or absent. That is, there may be *insignificant* insulin exposure or stimulating effect of the insulin-responsive cellular machinery. Consequently, the experiment design subspace sampled will be too small to adequately explore the expression space of interest. Furthermore, if the sample population is ethnically homogeneous or, in the case of a mouse experiment, shares the same genetic background, then these samples will only inform us of the reproducibility and noise of the biology and the measurement system. In contrast, if the same population were exposed to varying degrees of insulin, then the behavior of the genetic program in response to insulin might be better understood. Alternatively, taking a population known to have varying degrees of insulin resistance would also provide a more diverse range or heterogeneity of interactions in insulin signaling, which would enable the investigator to explore the space of possible relationships between the genes. This would then allow the application of a machine-learning algorithm to elucidate these dependencies. If the degree of insulin resistance is clinically measured (*e.g.*, as obtained in a carefully performed hyperinsulinemic euglycemic clamp[3]) in close temporal proximity to the expression measurements, then the investigator will probably have the most accurate control of the experiment design space. The goal then is to stress or stimulate the cellular regulatory system, which in this instance is glucose homeostasis and insulin signaling, so that the genes "show us" how they interoperate under normal, and pathological or pharmacological circumstances. There are several other obvious variations of these kinds of stimulations such as:

1. Examine the expression profile in adipose tissue at different points in time after the glucose exposure to understand the timing of the genetic program. We should not expect all genes to act at the same time or rate, and therefore sampling the changes in the transcriptome across multiple time points at the appropriate frequency will be much more revealing than just one or two points in time.

2. Assess gene expression at different fixed concentrations of insulin.

3. Assess gene expression in ethnically diverse populations. The difference in genetic backgrounds of these populations will give a sense of how much variation in the signaling may be due to genes outside the set that is being studied in a particular expression experiment. This implies that we would need to hypothesize that

---

[3]This involves giving the subject a continuous intravenous infusion of insulin and glucose, while maintaining a relatively constant blood glucose concentration by varying the rate of the glucose infusion.

changes in the genetic background affect the functioning and thus the functional genomics of the system. This presupposes that

- insulin production and insulin effect is the combined result of the activity of several genes;

- polymorphisms in these genes are distributed heterogeneously across human populations;

- such polymorphisms result in significant changes in gene expression or function, or both.

4. Explore the pharmacological effects of insulin. After a certain dosage level, increasing levels of insulin may no longer result in increased effects of glucose transport. However, increasing insulin may affect other physiological systems within the same cell, as, *e.g.*, transporting ions across the cell membrane. Taking tissue samples at pharmacological concentrations of insulin and comparing their expression profile to those obtained with insulin concentrations in the physiological range would allow determination of genetic mechanisms that respond to the pharmacological dosing.

In summary of our discussion, we note that an appropriate set of (microarray) experiments is that in which the cellular response to the stimuli involved in the experiments will exercise the interaction between genes over the range of interest and relevance to the biologist-investigator. This critical aspect of the experimental design will trump any subsequent choice of bioinformatic analytic technique. Conversely, *no* bioinformatics analysis will be sufficient to extract functional genomic knowledge if the expression space is not adequately explored.

2.1.4   Discarding data and low-hanging fruit

After an investigator obtains his or her first microarray data sets, often the first question he or she will ask is: "Which of the observed up- or downregulation of genes represents biologically significant changes in expression?" This and similar questions are addressed in detail in section 3.3. Here we provide a context and motivation for an answer that biologists often find discomfiting: that *very few genes are in fact significantly changed in expression in a way that is distinguishable from biological and measurement variation and noise.*[4] We should emphasize at this point that there is an important distinction between mathematical or statistical significance, and biological significance. The former has to do with analytically

---

[4]This is particularly true given the noisy nature of most microarray-based expression measurement systems.

quantifying the difference between two or more different sample sets of numbers, *i.e.*, microarray expression data, for which there exists standard statistical tests like the Student's t or χ^2. These tests typically ignore the rich biological meaning and structure in these numerical representations of expression. The latter is a more complex matter which we shall discuss here.

It is likely to be the case that there are thousands of genes with important biological outcomes effected by small relative changes in the expression levels. In the present state of the microarray technology and with the typically small number of replicate experiments for any given condition, these small changes cannot be reliably and reproducibly distinguished from noise. This may mean that even though we measure tens of thousands of genes in a microarray experiment, we would only obtain hundreds of genes that we are reasonably convinced are involved in a particular biological system. On the other hand, not all statistically significant gene expression changes lead to a significant change in the biological or physiological state of a system. This answer is most disconcerting to those biologist-researchers who have had substantial experience in investigating the expression of only a few genes at a time. Their investigations typically involve one or two highly likely candidates based on many prior investigations, and where multiple measurements (all of which are carefully checked) have been made of the expression levels. Consequently, the researchers have well-grounded ideas of what constitutes a biologically relevant or significant change in expression for these few genes. They do not have the benefit of this sort of knowledge for each of the thousands of genes that are represented on a microarray but nonetheless are uncomfortable with discarding from further analysis thousands of genes that appear to have large numerical changes in expression.

In contrast, due to the large number of genes that can be measured in a single experiment, computer scientists and computational biologists are comfortable—perhaps too comfortable—with generating exhaustive lists of genes that are possibly involved or interacting in a particular biological process. This kind of exhaustive list generation has been quite common in publications of microarray experiments from 1997 until today. Unfortunately, such lists are not that helpful to the biologists who wish to determine which elements in these lists are worth pursuing, *i.e.*, are biologically significant for their investigations. This is often the reason for questions of what constitutes a significant fold change, where significance can mean anything from "present in the tissue," "associated with the system being studied," "causative of the changes in the system being studied," or even simply meaning "worthy of further study." Thus, an analyst for a typical functional genomics study will simply draw two boundaries: one covering increases in expression and the other covering decreases in expression, as shown in figure 2.10. Such an analyst will then declare that any gene found beyond these lines (the red points) is significant or relevant to the biological process being studied. Unfortunately, there is no single threshold number or even function that we can provide that would generate a particular number of candidate genes that is worth pursuing for any arbitrary experiment and experimental design. Fortunately, there are well-founded decision analytic procedures that can be followed in order to come up with the appropriate number of candidate genes for any given investigation. These are described in detail in section 3.2. The underlying motivation for picking a threshold is straightforward

The first question one has to consider is the cost of a *false positive*. A *false-positive* gene is a gene which, in reality, is not biologically significant to the process under investigation, but which the analytic technique deems to be significant, statistically or biologically. That is, what is the cost involved in the follow-up procedure to confirm the function of a gene in a particular biology process? Whether this confirmation process, or biological validation, involves quantitative PCR, *in situ* hybridization, the generation of a transgenic mouse, a transient-transfection assay, or transfecting a cell line, its cost in time and money will be substantial. This cost thus limits the number of genes that can be investigated. The threshold that will then be picked will be determined in part by the disutility or cost of the biological validation step for false positives. The tens of thousands of genes present on any microarray now ensure that the number of false positives could potentially overwhelm the typical time and financial budgets of most (academic) research laboratories.

Of course, there is the second and converse question of the cost of missing a *false negative* from the same system. A *false-negative* gene is a gene which, in reality, is biologically significant to the process under investigation, but which the analytic technique deems not to be significant, statistically or biologically. There are likely to be several genes involved in any signaling pathway of interest. Not all genes in a pathway are equally amenable to biological validation. Furthermore, some of the chosen genes may be more suitable targets than others for diagnostic assays or therapeutic interventions. If the threshold picked for considering a gene to be significantly[5] changed in expression excludes one of these targets, then the cost of that false negative will be quite high, *e.g.*, a missed scientific opportunity or lost opportunity for commercial development.

With this in mind, it becomes clearer how a threshold should be picked. First, the investigators should ask themselves how many false-positive and false-negative leads they can tolerate. Then they can conduct a series of replicate experiments as

---

[5]For the discussion here, we define "significant" as "worthy of further study."

discussed in section 3.2. and determine where they will have to draw the thresholds to attain the required sensitivity and specificity.[6] So that if one is about to embark on a functional genomics investigation, an integral part of the experimental design must be the decision analysis diagrammed in figure 2.9. This requires that the cost to one or one's enterprise of missing the one or more genes that are likely to be involved in the genetic regulatory pathway of interest be made explicit. Similarly, the cost of having to follow up on false leads must be estimated. These two costs define the principal disutilities that one is trying to minimize. In order to complete this simple decision analysis, the probabilities corresponding to the sensitivities and specificities that are diagrammed will also have to be obtained. After this is complete, one will then know whether it is possible to engage in a productive high-throughput functional genomics strategy, or whether one has to increase the sensitivity or specificity of the measurement techniques, or whether one needs to increase one's biological validation budget, or all three. Given the current level of reproducibility of expression measurements in microarrays, most investigators will choose a high level of specificity to reduce the false-positive rate. This will inevitably lead to a high false-negative rate. At this point the biologist will find that among the false negatives are genes that are known to have changed expression (but did not meet the significance threshold computed by the bioinformatician) which will lead naturally to the following worry: Many other genes of relevance to the biological system under investigation are being discarded from the analysis and follow-up. This worry is likely to be well-founded but given the state of the measurement technology, and the typical sample sizes employed, it cannot be easily remedied. It remains, that the *low-hanging fruit*, while only numbering in the hundreds, are likely to shed new light on the processes studied. Already this represents several orders of magnitude more hypotheses to be tested as compared to the investigations of the pregenomic era. For those fruit higher up the tree, the above decision analysis suggests the need to wait for more accurate and cheaper microarray technologies.

It goes without saying that the preceding discussion is an oversimplified decision analytic procedure. Attempting to assess the utilities of false-negative findings will be particularly difficult for most investigators, but the constraints of realistic budgets will be a driving factor in considering this analysis. However, the procedure we have described does have the merit of providing a rational basis for deciding upon the number of genes that one should seek to obtain from one's pipeline. Even a coarse decision model can be useful in avoiding unpleasant surprises after the

---

[6]Sensitivity can be described as $(1 -$ false negatives$)$/all positives; specificity as $(1 -$ false positives$)$/all positives.

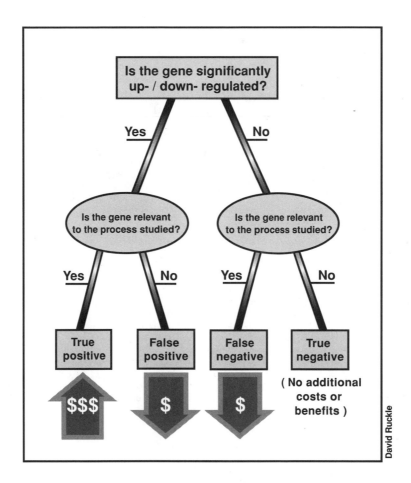

Figure 2.9
A decision analytic procedure for picking a threshold for selecting genes from a functional genomics experiment. Because of the large number of probes on current microarrays, it is all too easy to underestimate the cost and practical intractability caused by even a moderate false positive rate. Investigators are advised to perform this simple decision analysis in order to determine what false positive and false negative rate they can afford before they proceed with any experiments.

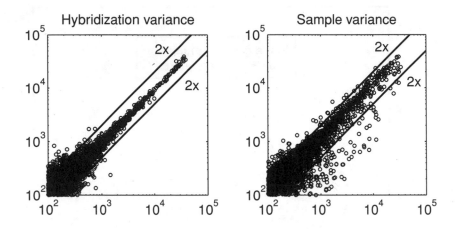

Figure 2.10

Typical uses of fixed fold thresholds. Reproducibility scatter plots. In each of the experiments, samples were hybridized to identical oligonucleotide arrays containing probes for 6800 human genes. The message abundance in arbitrary units is plotted. Left panel: A single biotinylated RNA target was divided in two and each half hybridized to two arrays. Fifty-nine genes (0.8%) judged to be expressed "present" by the Affymetrix GeneChip software differed by more than twofold, and 0.3% differed by more than threefold. Middle panel: A single biotinylated target was hybridized to one array, the sample removed and then rehybridized to a second array. Of "present" genes 2.6% and 0.4% differed by more than twofold and threefold, respectively. Right panel: A single total RNA sample was converted to biotinylated cRNA in two independent labeling reactions, and the cRNA then hybridized to two arrays. Of the "present" genes 2.2% and 0.4% differed by twofold and threefold, respectively.

initial excitement of obtaining a list of putatively interesting genes wears off. The reader who is interested in more sophisticated decision analyses is referred to the textbook by Weinstein *et al.*[187]. It should be noted that the raw specificities and sensitivities of the expression microarrays used in a functional genomics pipeline need not be the ones that are used in this decision analysis. As we demonstrate in sections 3.2.2 and 3.2.8, the application of appropriate noise models over repeated experiments can lead to improved sensitivity and specificity. The cost of such models is typically in the increased number of microarray experiments required to develop those models.

Dynamics   Perhaps the best glimpse of the interactions between the various genetic components, and in particular their causal dependencies, is obtained through analysis of the dynamics of the system. This is elaborated considerably in the section on dynamics (section 3.6.3, page 146). Suffice it to say here that understanding the trajectory over time of various RNA component species rather than their absolute values at any given point in time provides far more information about the operation of the underlying system. To return to our watch metaphor, observing how the gears turn inside the watch while we smoothly move the minute hand or the hour hand will give us much more information about how the gears move together than a few (unordered) snapshots in time of the different positions of these hands on the face of the watch. In most analyses to date, even when expression data are obtained as a time series, the information buried in the specific ordering and timing of these measurements is rarely exploited. For example, many of the clustering and classification methods employed (and introduced below) will generate the same answers even if the order of the measurements were shuffled. In fairness, the number of time points and the sampling intervals of many of the existing time-series expression data sets is minimal. They provide insufficient information for most of the standard armamentarium used to analyze time-oriented data. As the price of microarrays falls, the quantity and quality of time-series experiments will undoubtedly increase, at which point many of the techniques for analyzing trends and periodicity that have been developed in statistics and the signal processing literature will become applicable. At that time, the temporal relationships between every single expressed gene with every other gene measured at all time points will provide a qualitatively improved set of insights into causal relationships in the genetic regulation of cellular physiology.

2.2   Gene-Clustering Dogma

There is a central underlying assumption in all gene-clustering techniques for expression analysis. Simply put, the assumption is that genes that appear to be expressed in similar patterns are in fact mechanistically related. Furthermore, the corollary to this assumption is that although genes may distantly affect the function of other gene products, they fall into groups of more tightly regulated mechanisms.[7] For instance, the genes that govern chromosome function or meiosis may be more tightly linked to each other than they are to the genes involved with another function, such as apoptosis. This has been the basis of our collective experience in biological investigations over the last century: that there are groups of proteins which interact together more closely than others. Often they have been organized into pathways such as glycolysis, the Krebs cycle, and other metabolic pathways in which the gene products called enzymes have to work in concert. Other more obvious functional clusters are those of structural proteins which have to come together in a conserved and reproducible fashion in order to serve their purpose, whether they are the components of the ribosomal unit or the histoproteins which are essential to maintenance of the structure of chromatin. On this basis, if we can find genes whose expression patterns approximate one another, we can possibly impute that they are functionally clustered together. That is, they have functions that are related.

Several important caveats are worth noting here. First of all, it remains unclear just how discrete the functional groupings of gene function are in the cellular apparatus. It may be that individual gene products have so many different roles under different circumstances that several of them partake of essential roles in significantly different functions.[8] The second caveat is that the term "functionally related" is itself ill-specified. If the pattern of expression of one gene is similar to that of another, it could signify all kinds of relationships, ranging from "two genes having gene products that physically interact," to "one gene encoding a transcriptional factor for the other gene," to "two genes having different functions but similar promoter sequences," to "two genes both with promoter sequences bound by repressors which are knocked off when a nuclear receptor is activated, even though the two genes have widely disparate functions." Of course there is a level of abstraction at which *all* genes are functionally related in their role of keeping the cell alive and producing

---

[7]In other words, there are several distinct and statistically distinguishable processes.

[8]Well-known examples of these are the transcriptional factors such as sonic hedgehog (*shh*) which in some tissues and at some times are involved in cell proliferation and in others in cell differentiation processes.

whatever components are needed for the rest of the organism. But below this level of abstraction, there are many alternative and, by their nature, sloppy definitions of clustering. Therefore, we should be somewhat wary of the claim that similarity in expression corresponds to similarity in function. Nonetheless, it is a useful starting point for many analyses of a genome whose function remains by and large unknown at this time. Additionally, as we discuss in the chapter on dissimilarity measures (chapter 3), the question of what constitutes a similar expression pattern is itself poorly defined, or at least has multiple alternative definitions. For example, similarity could mean having similar patterns of change over time. It could mean similar absolute levels of expression at any given point in time, or it could mean perfectly opposite but well-choreographed patterns of expression. Just which dissimilarity measure is chosen for looking at patterns of expression will influence the kind of functional clusters that we expect.

2.2.1 Supervised versus unsupervised learning

In the introductory chapter, we discussed why the computational techniques applied in the analysis of gene expression are qualitatively different from those of traditional biostatistics: the data sets are of high dimensionality and yet the number of cases are relatively small. Consequently, the number of solutions that could explain the observed behavior is quite large. For this reason, the machine-learning community has recognized the potential role for their techniques specifically designed to explore high-dimensional spaces (such as those of voice or face recognition) and have also recognized the enormous need to apply these techniques to genomic data sets. To this effect, the first reaction of a computer scientist with a background in machine learning when he or she becomes aware of the new challenges created by genomic data sets is to pull out the tools of the standard armamentarium of machine learning. He or she then begins to explore informally, and subsequently evaluates formally the results obtained when these tools are applied to genomics data sets. We provide a framework for the armamentarium in the genomic data mining chapter (chapter 4, page 149) but here we only provide enough of an overview to discuss these tools with respect to experimental design.

Two useful broad categorizations of the techniques used by the machine-learning community are *supervised learning techniques* and *unsupervised learning techniques*. These are also commonly known as classification techniques and clustering techniques, respectively. The two techniques are easily distinguished by the presence of external labels of cases. For example, labeling a tissue as obtained from a case of acute myelogenous leukemia (AML) or acute lymphocytic leukemia (ALL) is needed first before applying a supervised learning technique to create a method to learn which combinations of variables predict or determine those labels. In an unsupervised learning task, such as finding those genes that are co-regulated across all the samples, the organization or clustering of the variables operates independently of any external labels. The kinds of variables (also known as *features* in the jargon of the machine-learning community) that characterize each *case* in a data set can be quite varied. Each case can include measures of clinical outcome, gene expression, gene sequence, drug exposure, proteomic measurements, or any other discrete or continuous variable believed to be of relevance to the case.

What kinds of questions are answered by the two types of machine learning? In supervised learning, the goal is typically to obtain a set of variables (*e.g.*, expressed genes as measured on a microarray) on the basis of which one can reliably make the diagnosis of the patient, predict future outcome, predict future response to pharmacological intervention, or categorize that patient or tissue or animal as part of a class of interest. In unsupervised learning, the typical application is to find either a completely novel cluster of genes with a putative common (but previously unknown) function, or more commonly to obtain a cluster or group of genes that appear to have similar patterns of expression to a gene (*i.e.*, they fall into the same cluster) already known to have an important well-defined function. The goal there is to find more details about the mechanism by which the known gene works and to find other genes involved in that same mechanism either to obtain a more complete view of a particular cellular physiology or, in the case of pharmacologically oriented research, other possible therapeutic targets. Although the distinct goals of supervised versus unsupervised machine-learning techniques may appear rather obvious, it is important to be aware of the implications for study design. For example, an analyst may be asked to find classifiers between two types of malignancy, as was done in the Golub *et al.* investigation of AML and ALL [78]. However, the lists of genes that reliably divide the two malignancies may have little to do with the actual pathophysiological causes of the two diseases and may not represent any particular close relationship of those genes and function. Why might this be? It is quite possible that small amounts of change of some gene products such as transcriptional activators and genes such as *p53* may cause large downstream changes in gene expression. That is, with only a subtle change, an important upstream gene may cause dramatic changes in the expression in several pathways that are functionally only distantly related but are highly influenced by the same upstream gene. When applying a classification algorithm directly on the gene expression levels, the algorithm will naturally identify those genes which change the most between the two or more states that are being classified. That is, a study design geared toward the application of a supervised learning technique may generate a useful artifact for classification, diagnosis, or even prognosis, but it will not necessarily lead to valuable insights into the biology underlying the classes obtained.

Let us consider the more general cases where gene expression values are not the only data type. For example, as illustrated in figure 2.11 below, a given case can include several thousand gene expression measurements but also several hundred phenotypic measurements such as blood pressure, a laboratory value, or the response to a chemotherapeutic agent. Here again a clustering algorithm can be used to find those features that are most tightly coupled in the observed data. When designing an experiment that includes the various data types, it is worthwhile thinking ahead of time whether some kinds of features are more likely to cluster together, separately from the genomic data. That is, after application of a clustering algorithm the data set may reveal relationships between the nongenomic variables that are much more significant and stronger than any of those that involve gene expression or sequence. While that is not necessarily a bad outcome, it will not help the investigator who is trying to understand the particular contribution of genetic regulation to the observed phenomenon. As an example, if one looks at the effect of thousands of drugs on several cancer cell lines, then it should not be surprising if these drug effects were most tightly clustered around groups of pharmaceutical agents that were derived from one another through combinatorial chemistry. Similarly, phenotypic features which are highly interdependent, such as height and weight, will cluster together. The strength of these obvious clusters will often dominate those of heterogeneous clusters that contain phenotypic measurements as well as gene expression measurements. This suggests that careful use of feature reduction to only include those features that are nonredundant and only truly independent phenotypic measures for each case should be used. We refer the reader interested in systematic approaches to feature reduction to the excellent text by Sholom Weiss and Nitin Indurkhya [188].

2.2.2   Figure of merit: The elusive gold standard in functional genomics

Whereas the discussion above was motivated by the questions posed by our collaborators who are primarily biologists or clinicians, it parallels similar questions coming from those colleagues with a computer science background. One question is, how can we determine whether a particular methodology that we are applying is successful or not? Is one machine-learning algorithm better than another, or does one clustering method provide more robust clusters than another? In other words, how do we know how successful a particular functional genomics investigation is? What is the *figure of merit* that we are trying to obtain?

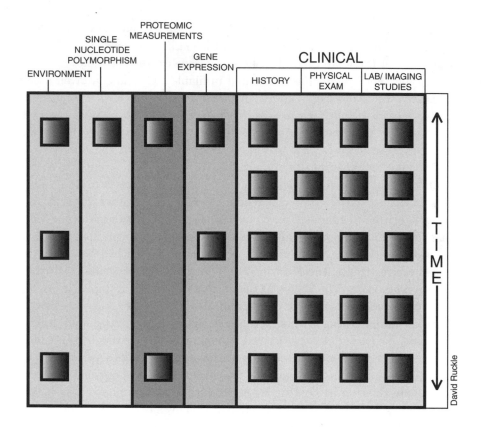

Figure 2.11
Clustering more than only expression values. A full-fledged functional genomic study with clinical relevance will involve multiple and quite heterogeneous data types. The realities of clinical research and clinical care will ensure that there is a requirement for the handling of missing data.

The figure of merit is probably the most ascertainable for the case of a known classification. Take, for instance, the case of the task of the classifying whether a particular tissue belongs to one of two tissue types using a supervised learning technique. Suppose also that a test sample is available. Then, various classification algorithms can be run against one another and the sensitivity and specificity of these algorithms can be compared. More generally, the receiver operating characteristic (ROC) curve of each of these classification algorithms can be defined. A good example of such a comparison is provided by Michael Brown *et al.* [33]. Several classification algorithms were employed and compared and ranked based on a weighted measure of true and false positives and true and false negatives. The best performance turned out to be in the application of support vector machines. The test that was used for this "bake-off" was the *Saccharomyces cerevisiae* data set from Stanford involving expression data of 2467 genes across 79 different hybridization experiments.[9] Because the classification is known in advance, the performance of these algorithms can be measured. However, there are several limitations which should be recognized. The classification algorithms' performance only pertains to the particular population or set of experiments on which it was originally tested. That is, a classification algorithm working on a yeast data set may not necessarily work as well relative to other algorithms on distinguishing two different types of leukemia or two different types of diabetes mellitus in humans. Even less dramatically, the same algorithm trained on one set of patients with leukemia may have a different relative performance compared to other classification algorithms on another set of leukemic patients drawn from a different population or obtained with a different ascertainment bias. For example, one set of patients may have been selected because of particularly refractory disease that caused them to be referred to a tertiary care hospital and another set of patients may have been treated in community hospitals. It is possible that the underlying diseases of these populations may be different and therefore the possibility of underfitting or overfitting the underlying classes of disease in these populations will differ. The figure of merit, therefore, in a classification experiment is the correct classification performance of an algorithm or measure for that particular population. Its applicability or merit for use in even related populations is uncertain or problematic.

---

[9] Although this paper is one of the better examples of comparison and classification techniques, it does not compare the entire ROC curve but only one single point on the curve. Therefore, it remains unclear what would be the overall performance of these algorithms under circumstances in which specificity and sensitivity were valued differently. See page 201 for a brief discussion of using the ROC curve to evaluate a classification test.

Figure of merit in a clustering experiment   What is the figure of merit for a clustering analysis? That is, how can we evaluate the degree to which the cluster of genes obtained are in fact correct or relevant? For anybody who has performed a clustering procedure on microarray expression data, it will seem obvious that it is altogether too tempting when one obtains a cluster of genes through any machine-learning algorithm to come up with a *post hoc* justification for why those genes may have fallen or risen together. This temptation is all too evident in the publications of the last 3 years in which clusters of genes reported by investigators are described typically as falling into one of three different categories. The first is well-known published associations, either mechanistically verified or empirically known, that have been obtained through other more conventional techniques. Second is associations that appear plausible to the authors, and presumably the reviewers, but have not been proven in the literature. The third is associations for which the authors are hard-pressed to find support. For the biological investigator who wishes to further investigate the fundamental biology of genes and their function, the last two categories constitute a fairly unsatisfactory basis on which to invest several months, if not years, understanding the basis for the imputed associations or potential functional dependencies.

The challenge then for the functional genomist using microarray data is to come up with means to validate the clusters obtained. There are two levels of validation. The first is a statistical or methodological validation with techniques such as permutation or cross-validation as described elsewhere in this book (section 4.12.1 and section 4.12.2). These techniques can be used to ensure, within a specified degree of certainty, that for the particular data set being studied, the clusters are neither the result of serendipitous coordination of gene behavior nor that the samples are apparently inadequate[10] to estimate the reliable coordinated behavior of gene expression. An alternative method is to estimate the probability of particular clusters. This methodology is still only in its infancy and the relevant literature is only now beginning to be generated for functional genomics.[15, 178]. As with other aspects of functional genomics, much larger data sets will allow more effective use of these methods. Ultimately however, we should remember that the clusters or classifications that we obtain through machine-learning techniques are only reflections on the measurements made in a particular system. If we are to make broad claims about the empirical or scientific truth of the relationships or classes that we infer from expression data, we will have to submit analyses to the same kind of

---

[10]Insufficient numbers of biological samples, insufficiently strong effects, or sufficiently distinct biological processes, or any combination of these.

1. It (the suspected pathogen) should be present in every instance of the disease.

2. It should be isolated from the diseased host and grown in pure culture.

3. The specific disease must be reproduced when a pure culture of the agent is innoculated into a healthy susceptible host.

4. The same agent must be recoverable from the experimentally infected host.

Table 2.2
Koch's postulates

tests that have been developed for other experimental sciences with longer histories. Specifically, we need to at least come up with the microarray equivalent of Koch's postulates. In 1890, Robert Koch set out to develop criteria for judging whether a given bacterium was the cause of the disease. The criteria are summarized in table 2.2.

It has been recognized there are several problems with Koch's postulates and there is a great deal of discussion and disagreement about this in the literature of the history of science [143]. However, the postulates are a good first approximation of the requirement for validation of functional genomics experiments. Therefore the analogs to Koch's postulate in the domain of microarrays may be useful. These are illustrated in table 2.3. The criteria describe the test for an inferred regulatory dependency obtained through microarray analysis.

It is only with these biological tests that we can "rate" the performance of clustering algorithms—it is the only durable figure of merit. It is not surprising that very little of this kind of hypothesis testing has been reported in the literature. After all, for each gene in a cluster, there is implied a laborious sleuthing effort in which the bioinformatician and biologist must closely collaborate to verify the results using the analogs to Koch's postulates in table 2.3. If we are to avoid a functional genomics "meltdown," hinted at in the introductory chapter, then the interdisciplinary nature of the functional genomics pipeline (diagrammed in figure 1.3.1) will have to be substantively adopted. All gene associations or regulatory relationships suggested by bioinformatics techniques should undergo at least a minimal validation by the biologist. Conversely, experimental design without the involvement of the

1. If gene A is found to be correlated with gene B then this relationship should be reproducible through northern blots and other quantitative expression measurement techniques [151].

2. Furthermore, in all pathological conditions in which the gene is thought to play a critical role, the predicted level of that gene should be found at the right time and in the right part of the tissue during the disease process.

3. If a gene thought to be involved in a pathway is underexpressed or overexpressed in a model system, then the process controlled by that pathway will be affected.

Table 2.3
Functional genomics analogs to Koch's postulates

bioinformatician is likely to lead to wasteful, expensive, and unrewarding analyses.

3 Microarray Measurements to Analyses

3.1 Generic Features of Microarray Technologies

The gene expression assaying technologies that will be discussed in this chapter are RNA detection microarrays, variously known as DNA chips, biochips, or simply chips. It is easiest to think of the utility of microarrays in functional genomics as a five-step process.[1]

- *Probe*: This is the biochemical agent that finds or complements a specific sequence of DNA, RNA, or protein from a test sample [8]. It can include pieces of cDNA amplified from a vector stored within a bacterial clone, oligonucleotides synthesized in a fluid medium, oligonucleotides built on a solid phase base pair by base pair, or nucleotide fragments of a chromosome. This can be extended to incorporate the database of expressed sequence tags (ESTs) for serial analysis of gene expression (SAGE), (see Glossary, page 277).

- *Array*: The method for placing the probes on a medium or platform. Current techniques include robotic spotting, electric guidance, photolithography, piezoelectricity, fiber optics, and microbeads. This step also specifies the type of medium involved, such as glass slides, nylon meshes, silicon, nitrocellulose, membranes, gels, and beads.

- *Sample probe*: The mechanism for preparing RNA from test samples. Total RNA may be used, or mRNA may be selected using a polydeoxythymidine (poly-dT) to bind the polyadenine (poly-A) tail. Alternatively, mRNA may be copied into cDNA, using labeled nucleotides or biotinylated nucleotides.

- *Assay*: How is the signal of gene expression being transduced into something more easily measurable? For the microarrays in common use, gene expression is transduced into hybridization. For SAGE, gene expression is transduced into oligonucleotides via restriction enzymes and ligation. For PCR, gene expression is transduced into amplified pieces of cDNA.

- *Readout*: How is the transduced signal going to be measured? How is information about the signal represented? For the microarrays in common use, hybridization is typically measured either using one or two colored dyes, or radioactive labels. For SAGE, the constructed oligonucleotides are measured through sequencing. For PCR, the amplified pieces of cDNA can be measured using gel electrophoresis.

---

[1]This schema is adapted from Leming Shi's website `http://www.gene-chips.com`.

For the microarrays in common use, one typically starts by taking a specific biological tissue or system of interest, extracting its mRNA, and making a fluorescence-tagged cDNA copy of this mRNA. This tagged cDNA copy, typically called the *sample probe*, is then hybridized to a slide containing a grid or array of single-stranded cDNAs called *probes* which have been built or placed in specific locations on this grid (see table 3.1 for alternative terminology). Similar to the general hybridization principles behind Southern or Northern blots (see glossary, page 277), a *sample probe* will only hybridize with its complementary *probe*. Fluorescence is typically added to the sample probe in one of two ways: Either (1) fluorescent nucleotide bases are used when making the cDNA copy of the RNA, or (2) biotinylated nucleotides are first incorporated, followed by an application of fluorescence-labeled streptavidin, which will bind to the biotin. Depending upon manufacturer-specific protocols, the probe-sample probe hybridization process on a microarray typically occurs over several hours. All unhybridized sample probes are then washed off and the microarray is lit under laser light and scanned using laser confocal microscopy analogous to the phosphor imager in the traditional blot procedures. A digital image scanner records the brightness level at each grid location on the microarray corresponding to particular RNA species.

Studies have demonstrated that the brightness level is correlated with the absolute amount of RNA in the original sample, and by extension, the expression level of the gene associated with this RNA [157]. At a coarse level, one microarray experiment may be thought of as N many Northern blots that simultaneously assay a total RNA sample on a small common medium or substrate for as many different mRNA species—N being the total number of unique RNA species probes located on the physical chip. Note that the amount of total RNA required for a typical Northern blot is more than sufficient for one microarray experiment in the current technologies. However, this analogy breaks down in that only a single hybridization condition (*e.g.*, temperature, time) is used in hybridizing all N assays, and unless the probes are carefully chosen, this may not be the optimal condition for the assay of all RNA species.

A prominent characteristic of microarray technologies is that they enable the comprehensive measurement of the expression level of many genes simultaneously on a common substrate. In this regard, typical applications of microarrays include the comprehensive quantification of RNA expression profiles of a system under different experimental conditions, or expression profile comparisons of two systems under one or several conditions. The former embraces the comparison of expression profiles of a system under a control and a test condition, whereas the latter includes contrasts between different strains of organisms, as, for instance, between a normal

	Immobilized nucleic acid on microarray surface	Free nucleic acid that is being interrogated
General microarray	Probe	Sample Probe
Robotically spotted microarray	Probe	Probe
Affymetrix microarray	Probe	Target

Table 3.1
Common probe nomenclature

(*e.g.*, wild-type), and a constructed (*e.g.*, knockout) organism. Another intriguing use of microarrays is to compare expression levels between neighboring cells within the same microscopic field, as demonstrated in [123]. Aside from their widespread utility in functional genomics, oligonucleotide microarrays have also been used for single nucleotide polymorphism (SNP) analysis since many probe SNPs can be placed on the microarray for comprehensive parallel SNP detection, and in much the same way one can also perform DNA sequence analysis.

Regardless of the gene expression technology to be adopted, almost all of them have performance factors that depend critically on the general validity of certain fundamental biological assumptions outlined below:

1. *There is a close correspondence between mRNA transcription and its associated protein translation.* As noted by Brown and Botstein [34], one would ideally like to measure the final products of every gene, such as proteins, or even better, the biochemical activity of these products, which are more directly related to biological functionality. Such quantitation would provide a link between chemical DNA bases at microscopic levels with biological aspects that are manifest at macroscopic scales such as phenotype and physiology. However, there is no practical generic tool to do this yet. The assumption that there is a principle of parsimony—akin to Hamilton's principle of least action in the physical sciences—which drives the close relationship between gene expression and biological function was most clearly articulated by P. O. Brown and D. Botstein in [34]:

> "The second reason [for using DNA microarrays to study gene expression on a genomic scale] is the tight connection between the function of a gene product and its expression pattern. As a rule, each gene is expressed in the specific cells

and under the specific conditions in which its product makes a contribution to fitness. Just as natural selection has precisely tuned the biochemical properties of the gene product, so it has tuned the regulatory properties that govern when and where the product is made and in what quantity. The logic of natural selection, as well as experimental evidence, provides part of the basis for our belief that there is a sensible link between the expression pattern and the function of its gene product. Thirty years of molecular biology have provided numerous examples of genes that function under specific conditions and whose expression is tightly restricted to those conditions."

Biologists can quickly find exceptions to this assumption. For example, proteins that make up the cellular matrix can considerably outlast the lifetime of their associated mRNA, or conversely, they may be metabolized or degraded much more rapidly than the mRNA from which they were transcribed. Nevertheless, the initial successes in the applications of gene expression microarrays in investigations of expression and function suggest that this assumption holds true more frequently than not.

2. *All mRNA transcripts have identical lifespans.* Again, there are several well-known exceptions. For instance, we know that length of the 3' poly-A tail of an mRNA appears to be related to its stability. Furthermore, there are examples of mRNA that have longer- or shorter-term stability within specific cells or after its transcriptional event such as dystrophin mRNA from patients with certain types of muscular dystrophies. As a tangential comment on temporal effects: The probe-sample probe hybridization rate is known to be a function of the guanine-cytosine (GC) content of a transcript. In general, this rate is proportional to GC richness.

3. *All cellular activities and responses are entirely programmed by transcriptional events.* At a meta-systems level, this assumption may indeed be true, but in terms of direct mechanistic coupling, there exist many examples in which external stimuli cause changes in the biochemical program within the cell *without* engaging the transcriptional machinery. Figure 3.1 is an illustration of a response of free intracellular calcium to aldosterone exposure. Aldosterone is a steroid hormone that typically acts through binding with receptors that are translocated to the nucleus and which then initiate or modify a transcriptional program. Here, the time scale of the acute response suggests a nongenomic mechanism; in other words, the response does not require transcriptional activation. This example demonstrates how a molecule that usually works via modulation of transcription (*i.e.*, steroid hormones) may also affect bioprocesses at the nongenomic level. There is also a much larger class of bi-

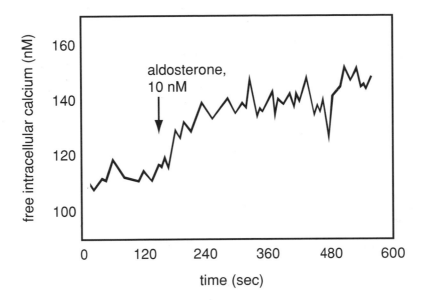

Figure 3.1
The nongenomic time scale response in aldosterone exposure. The response (in seconds) shown here is *much* faster than any known receptor-to-transcription response that steroid hormones are usually thought to act through. (Derived from Gamarra *et al.* [74].)

ological processes that do not primarily operate at the transcriptional level. These include muscular contraction, nerve excitation, and hormonal release. Eventually, all these events will cause some change in transcriptional activity, *e.g.*, replenishing stores of neurotransmitter, but the patterns of gene expression would probably not reveal the *control processes* that govern them at the subgenomic time scale.

There are presently two types of chip technologies in common usage: *robotically spotted* and *oligonucleotide* microarrays. It is important to note that these two microarray technologies are the very earliest and are currently the most widely used. Several competing technologies are emerging that may prove to be more cost-effective, reliable, and versatile (see chapter 7). Next, we briefly describe these two most popular microarray technologies.

3.1.1   Robotically spotted microarrays

Robotically spotted microarrays (figure 3.2), namely, the robotically spotted cDNA glass slide, were introduced into common use at Stanford University and first described by Mark Schena *et al.* in 1995 [157]. They are also known as cDNA microarrays.

In making the array, a robotic spotter mechanically picks up specific cDNA sequences, typically amplified from vectors in bacterial clones using PCR, from separate physical containers and deposits them in specific locations in the grid on the glass slide to create specific probes. Each cDNA drop should ideally be equal in quantity. This fabrication approach epitomizes the do-it-yourself tendency in microarray measurement, even though there are several commercial ready-to-use versions available. The frequently home-grown quality of these arrays has led to to the production of highly localized and customized microarrays which pose specific challenges in the dual tasks of background noise reduction and foreground RNA signal amplification during subsequent data analysis stages. These challenges are discussed in section 3.2. Designer-definable parameters exist to control spot-basing size, the amount of time for drying, as well as experimental parameters concerning the glass slide material to be used.

There are several advantages and disadvantages in using a robotically spotted microarray:

- The first advantage is customizability. A large subsequence ($\sim 2 \times 10^3$ base pairs long) complementary to the actual sequence that is to be probed is laid down on the chip by the designer who has full control over the species of probes that are to be used. This means, *e.g.*, that if one wishes to make customized chips for a specific purpose, such as for probing the expression of specific RNA in certain cell types, or specific polymorphisms of those genes, one can design a layout and can direct a robotic spotter to make these microarrays. The setup cost of this approach is on the order of $20,000 and detailed guidelines may be found at the Brown laboratory microarraying website.[2] Additionally, companies such as Affymetrix, Amersham Pharmacia, BioRobotics, and Cartesian Technologies sell robotic spotters-arrayer units. Glass slides are available from a variety of vendors specifically for microarray construction. Commercial aspects of the manufacture of such chips appear to be expanding. The notable disadvantage with greater customizability is that it may lead to more possibilities for errors. For instance, poor quality control or nonuniformity in the construction of different species probes such as spot-basing size will complicate the subsequent analysis and interpretation of the resulting chip data.

- The second advantage of robotically spotted microarrays is that larger pieces, or entire cDNAs, are placed on the chip, thus reducing the likelihood of nonspecific

---

[2]http://cmgm.stanford.edu/pbrown/mguide/index.html.

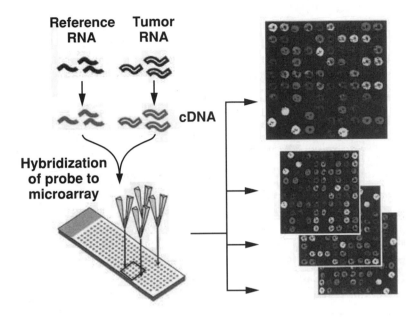

Figure 3.2
Robotically spotted microarray hybridized to two samples, each stained with two colored dyes.
An overview of procedures for preparing and analyzing cDNA microarrays and tumor tissue.
Reference RNA and tumor RNA are labeled by reverse transcription with different fluorescent
dyes (green for the reference cells and red for the tumor cells) and hybridized to a cDNA
microarray containing robotically printed cDNA clones. The slides are scanned with a confocal
laser scanning microscope, and color images are generated for each hybridization with RNA from
the tumor and reference cells. Genes upregulated in the tumors appear red, whereas those with
decreased expression appear green. Genes with similar levels of expression in the two samples
appear yellow. Genes of interest are selected on the basis of the differences in the level of
expression by known tumor classes (*e.g.*, BRCA1 mutation-positive and BRCA2
mutation-positive). Bioinformatics analysis determines whether these differences in gene
expression profiles are greater than would be expected by chance. (Derived from Hedenfalk *et al.*
[90].)

hybridization of labeled sample probes to the probe that was laid on the chip. Typically, a designer who wishes to have a particular gene probe on a chip will create a clone with the 5' and 3' ends of that particular gene of interest. The substrings of these cDNAs of the original gene are on the order of 100 to 200 base pairs long. A distinct weakness here is that even though a long probing subsequence ensures a sufficiently confident representative substring of the original gene, it does not mean that hybridization conditions will be fully and equally optimized for all species of cDNA subsequences. As we have noted earlier, the probe-sample probe hybridization rate is known to vary depending upon the GC content of a transcript.

• The third advantage is that RNA from two different samples (typically a test and a control condition) can be hybridized onto a common cDNA microarray substrate at the same time. The two separate RNA samples are typically labeled with different fluorescent dyes, such as Cy3 and Cy5. The two-dye system allows for the excitation of the microarray by laser light at two different frequencies and thus the image of hybridized RNA abundance can be scanned for both colors (corresponding to distinct samples) separately. Since the hybridization conditions, and thus the brightness, of any one spot is not the same as another spot, the individual signals are not typically used separately. Instead, the calculated signal is the *ratio* or *fold* difference in the brightness of the hybridized RNA of one sample versus another, specifically the intensity of Cy3 versus Cy5, *i.e.*, Cy3/Cy5. If a background intensity for each color is measured and controlled for, the ratio becomes $\frac{\text{Cy3-Cy3 background}}{\text{Cy5-Cy5 background}}$. This in turn provides a measure of relative abundance of the RNA from one sample with respect to the RNA from another sample, and hence the relative expression level of the corresponding gene in the samples. On the other hand, it has been shown that not all cDNA sample probe sequences label symmetrically with Cy3 and Cy5 [106] and paired dye-swapping experiments are performed (*i.e.*, if the first hybridization is control (Cy3) versus test (Cy5), then switch dyes for the second hybridization). If the labeling were symmetrical, then the plot of $\left(\frac{\text{Cy3}}{\text{Cy5}}\right)_{\text{Hybrization 1}}$ versus $\left(\frac{\text{Cy3}}{\text{Cy5}}\right)^{-1}_{\text{Hybridization 2}}$ for every probe would be line of slope one through the origin.

   In summary, each robotically spotted microarray experiment has its own built-in control, and results are given in terms of fold change or difference from a control situation. However, the measurement of *absolute* quantities and variations are more challenging in this technology.

3.1.2   Oligonucleotide microarrays

The second popular class of microarrays in use has been most notably developed and marketed by Affymetrix. Currently, over 1.5×10^5 oligonucleotides of length 25 base pairs each, called *25-mers*, can be placed on an array. These oligonucleotide chips, or oligochips, are constructed using a photolithographic masking technique similar to the process that is used in microelectronics and integrated circuits fabrication, first described by Stephen Fodor *et al.* in 1991 [70]. Currently, these commercially available microarrays are not produced individually, but instead are made in parallel. An entire wafer (containing between 40 and 400 microarrays) is constructed, tested, then broken apart to create the individual microarrays. At this time, commercially produced oligochips exist that are disease- as well as species-specific, such as rat neurobiology and yeast genome arrays, and custom microarrays can be ordered with a 4-week turnaround or less.

The manufacturing technique for an Affymetrix oligochip is markedly different from the more mechanical process of making robotically spotted arrays. Each wafer starts out as an empty glass slide. On this substrate, 25-mer probes are built base-by-base by placing single DNA bases on the glass and then on top of a preceding base. All 25-mer probes are constructed in parallel, with high precision, by selectively masking specific coordinates or locations on the glass surface and exposing the entire ensemble to ultraviolet light in between laying on the additional bases adenine (A), thymine (T), guanine (G), and cytosine (C) separately. Each applied photolithographic mask generates different areas of photodeprotection on the solid glass substrate. The combination of these masks with an intervening chemical coupling step allows the incorporation of additional nucleotides to existing strands only where desired. This entire process is known as *light-directed oligonucleotide synthesis.*

One advantage of the oligonucleotide microarray is that its higher density of probe pairs allows for more genes to be screened or assayed on a single chip, as compared to robotically spotted arrays. Consequently, there is less need for *a priori* restrictions on the number of genes that are to be scanned. A disadvantage is that the current technology only allows for at most one experiment to be run on a single chip at any one time. Thus, for example, one does not obtain meaningful data from placing the control sample and test probes on an oligonucleotide microarray simultaneously; instead these two samples are measured on two separate oligochips. This means, in turn, that one typically has to apply a suitable normalization transformation across separate microarray data sets (*i.e., inter-*array) at the subsequent data analysis stage in order to make meaningful comparisons of reported expression changes from a control to a test condition. Additionally, if a scientist is interested in studying a specific species for which no appropriate oligochip exists, then this technology is not presently available. Oligochips are not as easily customizable at the user's end as robotically spotted microarrays.

Since each probe is limited to 25 base pairs in length, a question immediately arises as to how each gene can be screened uniquely using only 25 base pairs. For each gene that needs to be represented, or whose expression needs to be measured on Affymetrix's oligochip, a set of sixteen to twenty 25-mers are chosen which uniquely represent that particular gene, and would hybridize under the same general conditions.[3] The Affymetrix literature calls the sample probe that is to be interrogated by cDNA probes on its microarray, the *target*. Every set of *perfect match* (PM) probes for an mRNA has a corresponding set of *mismatch* (MM) probes. An MM probe is constructed from the same nucleotide sequence as its PM probe partner, except that the middle (usually the 13th) base pair has been switched to result in an alphabet mismatch. For example, the following two 25-mers may be associated PM-MM probes to assay for the following sample probe:

Probe on chip $\left\{ \begin{array}{lll} \text{PM} & : & \text{ATCGACTGATGC}A\text{TGCATCCATCAT} \\ \text{MM} & : & \text{ATCGACTGATGC}C\text{TGCATCCATCAT} \end{array} \right.$

Sample Probe   TAGCTGACTACG T ACGTAGGTAGTA

The combination of a PM and its associated MM oligonucleotide probe is called a *probe pair*. There are two principal reasons for the use of MM probes. First, at low concentrations of the target or sample probe when the PM probes have already reached their lower limit of sensitivity, the MM probes display greater sensitivity to changes in concentration. Second, MM probes are thought to bind to nonspecific sequences at the same rate as the PM probes. Thus, MM probes serve as an internal control for *background* nonspecific hybridization (see section 3.1.2 for details). However, depending upon the total RNA sample, it could turn out that the PM probe is already highly specific and the MM probes are simply binding to differently-specific labeled subsequences in the sample. One has to be careful to distinguish between *nonspecific* hybridization and *differently-specific* hybridization with regard to the use of MM probe data.

For target preparation, sufficient amounts of sample probe are first synthesized by reverse-transcribing the total RNA using an oligo-dT primer containing a T7

---

[3]Early reports at the time of writing indicate the next generation of Affymetrix oligochips may use 11 probe pairs per probe set.

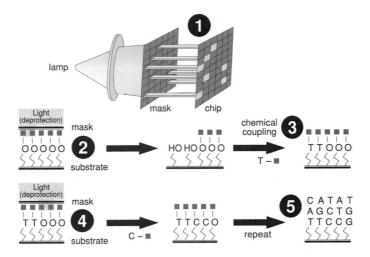

Figure 3.3
The photolithographic construction of microarrays. Synthesized high-density oligonucleotide microarray manufacturing with photolithography. Using selective masks, photolabile protecting groups are light-activated for DNA synthesis (*1, 2*); photoprotected DNA bases are added and coupled to the intended coordinates (*3*). This cycle is repeated (*4*) with the appropriate masks to allow for controlled parallel synthesis of oligonucleotide chains in all coordinates on the array (*5*). (Derived from Lipshutz *et al.* [121].)

polymerase site for 5' to 3' transcription. Amplification and labeling of the cDNA sample probe is achieved by carrying out an *in vitro* transcription reaction in the presence of biotinylated deoxynucleotide triphosphates (dNTP), resulting in the linear amplification of the cDNA population (approximately 30- to 100-fold). This linearity assumption becomes increasingly weak with decreasing quantities of total RNA and increasing number of amplification cycles. The biotin-labeled cRNA probe generated from the sample is then hybridized to the oligonucleotide arrays, followed by binding to a streptavidin-conjugated fluorescent marker. Laser excitation of the hybridized sample, confocal microscopy, and image acquisition by an optical scanner is performed. This results in an image file in which each oligonucleotide species is represented by a small rectangular area ($\sim 50\ \mu\mathrm{m}^2$), called the *probe cell*, which is itself composed of several image pixels, each occupying an area from 3 to 24 $\mu\mathrm{m}^2$ (figure 3.4). The image file is processed so that the intensity of each probe cell is reported as the 75th percentile of the intensities of all the pixels in a probe cell, excluding the pixels at the border of the cell. These probe cell intensities are stored in a `.cel` file, which therefore reports a measure of hybridization per contiguous oligonucleotide surface on the microarray. We describe the contents of a typical `.cel` file below.

Affymetrix microarrays and background intensity calculations   The Affymetrix GeneChip analysis protocol uses the term *absolute analysis* to describe the algorithm for determining whether transcripts represented on the probe array are detected and the intensity of expression. Briefly, for *each* microarray that has been hybridized with a prepared target, washed, and scanned, one obtains the following sequence of files as the Affymetrix software attempts to translate that particular chip experiment into numerical data representing detectable RNA intensity levels:

- A grey-scale `.tiff` image file of the physical microarray where a lighter or darker pixel indicates a respectively stronger or weaker hybridization of a cDNA fragment to the probes at the particular coordinate marked by the pixel

- A `.dat` text file containing coordinates and intensity levels for individual pixels

- A `.cel` text file with probe cell coordinates and intensity calculated as a trimmed average of pixel intensities—on average a probe cell is made up of 8 pixels × 8 pixels

- A `.chp` text file with 11 columns that contain the statistic, calculated by the Affymetrix software, for sets of 16 to 20 probe cell pairs, PM and MM, which interrogate particular transcripts

An important consideration in all microarray experiments is the contribution of the intensity of the *background effects* on each microarray. Since this can vary from one array to another, Affymetrix has developed its own methodology for subtracting the *background* from the hybridization signals. *Background effects* refer to brightness that ends up in the reported measurement reading for a probe cell, even though these effects did not originate from the probe cell. It typically includes *intra*-chip phenomena such as localized physical changes (*e.g.*, temperature during hybridization) in a probe cell that diffuse into its neighboring probe cells, or nonuniformity in ambient brightness levels in localized regions on a microarray surface. Background effects are components of the more general phenomenon of *noise*, which we discuss in section 3.2.

The entire microarray surface is divided into 16 sectors Within each sector an average statistic is calculated from the lowest 2% of probe cell intensity values. This is the background intensity for that sector and this value will be subtracted from the average intensities of all image features in that sector. Consequently, the number of background probe cells (*i.e.*, probe cells that are used to calculate background intensity) will depend upon the number of probe cells in the array. In theory, calculating background on a per-sector basis minimizes the effect of changes in the microenvironment across different parts of the array.

In addition to calculating average background intensity, background probe cells are also used to compute the effects of background *noise* variations on the reported measurements. As shown in figure 3.5, the mean intensity of these background probe cells is obtained and the distribution around the mean is calculated. In other words, it is assumed that all the background probe cells should be the same near-zero intensity, and the variation around this is considered noise. Intuitively, a wider distribution around the mean for the background probe cells implies a more pronounced noise component for all expression measurements in the probe array.

The calculation corresponding to this intuition is given in figure 3.5(b). This is the formula that Affymetrix uses to calculate the significance of a difference in intensities between the PM probe cells and associated MM probe cells. Both the ratio and the difference between the PM and the MM probes, PM/MM and PM-MM, for each probe pair are computed. These values are then compared against two thresholds respectively: the *statistical ratio threshold* (SRT) and the *statistical difference threshold* (SDT)— see figure 3.5C—which are themselves functions of the background for that probe array. If both thresholds are exceeded in the negative direction, the probe pair is considered "negative," and if they are exceeded in the positive direction, the probe pair is "positive." These assignments of probe pairs as negative or positive are then incorporated into more aggregate measures described

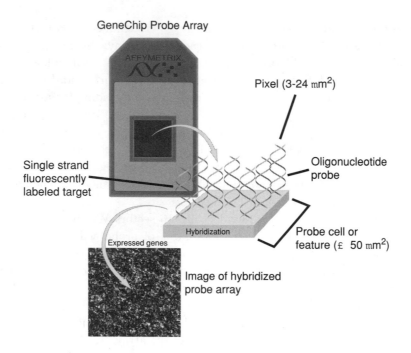

Figure 3.4
Pixel, probe cell, and Affymetrix scanned image. Schematic showing how images are composed
of probe cells, which contain probes that appear as pixels. (Derived from Jain [102].)

below.

For each gene whose expression the microarray has been designed to measure,
there are between 16 and 20 probe cells representing PM probes and a same number
of cells representing their associated MM probes. Collectively, these 32 to 40 probe
cells are known as a *probe set*. A .cel file contains all the probe cell intensities
for all the probe sets represented on a microarray. The .cel file is used, in turn,
to generate derived or aggregate statistics for each probe set (*e.g.*, a measure of
expression or a particular gene). These aggregate statistics are stored in a .chp
file.

Theoretically, each Affymetrix probe set of 16 to 20 25-mers is designed to be
uniquely representative of a particular gene or EST, and to no other known gene or
EST. However, it is often the case that a probe set cannot be found for a particular
EST which fulfills these rules, and thus must be designed by relaxing the rules. In
practice, nonspecific hybridization of individual 25-mers is more significant than in

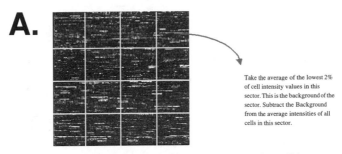

Take the average of the lowest 2% of cell intensity values in this sector. This is the background of the sector. Subtract the Background from the average intensities of all cells in this sector.

The number of **Background Cells** depends on the number of probe cells in the array:

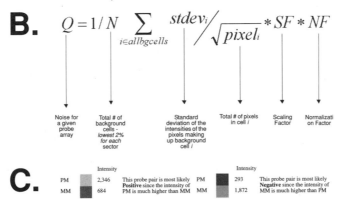

B.

$$Q = 1/N \sum_{i \in allbgcells} \frac{stdev_i}{\sqrt{pixel_i}} * SF * NF$$

Noise for a given probe array

Total # of background cells - *lowest 2% for each sector*

Standard deviation of the intensities of the pixels making up background cell *i*

Total # of pixels in cell *i*

Scaling Factor

Normalizati on Factor

C.

	Intensity			Intensity	
PM		2,346	This probe pair is most likely **Positive** since the intensity of PM is much higher than MM	PM	293	This probe pair is most likely **Negative** since the intensity of MM is much higher than PM
MM		684		MM	1,872	

The significance is determined by calculating both the ratio (PM / MM) and the difference (PM - MM) associated with each probe pair. These values are then compared against two threshold values: the **Statistical Difference Threshold** (SDT) and the **Statistical Ratio Threshold** (SRT). This is expressed mathematically as follows:

A probe pair is **Positive** if:

1. PM - MM † SDT

And

2. PM / MM † SRT

A probe pair is **Negative** if:

1. MM - PM † SDT

And

2. MM - PM † SRT

Note: not all probe pairs will be scored as Positive or Negative.

Figure 3.5
Background noise on Affymetrix arrays. **A**, A diagram showing how background noise variance is obtained from the background cells. **B**, The calculation of the background noise. **C**, The definition of SDT and SRT and how they define "positive" and "negative" probe pairs. (Derived from [10].)

Figure 3.6
Nonspecific hybridization on an oligonucleotide microarray. The two rows of probe cells
represent the probe set for a gene transcript. The PM probes are on the top row and the MM
probes are on the bottom row. Even if there is a lot of specific hybridization to the PM
oligonucleotides due to the presence of the targeted gene transcript, if there is significant
nonspecific hybridization, then the amount of transcript cannot be estimated accurately. In this
illustration, the number and intensity of dark probe cells on the bottom row is higher than that
in the top row and therefore most software packages would report that the reported intensity for
the gene is unreliable—an "Absent" *Absolute Call* in Affymetrix parlance.

robotically spotted microarrays, and this is one form of *noise* (see section 3.2 for
more coverage of noise). In addition, nonspecific hybridization with other RNA
species may result in an entire probe set for a gene fluorescing with little or no
noticeable contrast between the PM and MM, as shown in figure 3.6:

Recall that in theory, an MM probe is designed to bind to nonspecific sequences
at the same rate as its associated PM probe. Due to the low specificity of individual
oligonucleotide probes as compared to the lengthy cDNA sequences used in most
robotically spotted microarrays, Affymetrix's goal has been to use all 32 to 40 probe
cells to increase the aggregate specificity. In order to meet this goal, Affymetrix has
developed several measures across the PM and MM to increase specificity and generally improve the signal-to-noise ratio of the hybridization measurements. These
derived values provide measures of quantified gene expression and the reliability of
gene expression. These measures are summarized in table 3.2.

The two most commonly quoted measures are the *Average Difference* (Avg Diff)
and the *Absolute Call* (Abs Call). The Avg Diff is an aggregate measure of the
difference between the PM and MM probe cell intensities per probe set. In the
simplest case, a highly specific hybridization will "light up" all the PM probe cells
and none of the MM probe cells. In this case, the Avg Diff would be a positive
number which increases with the quantity of that particular RNA species present
in the sample. In practice, the pattern of figure 3.6 is more common. The greater
the intensity of the MM probe cells, the lower the Avg Diff. If the intensities of the
MM probe cells exceed those of the PM probe cells, then a negative Avg Diff can
be reported. Therefore, even if a gene is expressed highly in a particular biological
sample, the Avg Diff could be negative if there is also a lot of differently-specific
hybridization reported by the MM probe cells.

The various techniques in the literature that have been employed to handle negative values of Avg Diff in an analysis are all controversial. Most often, genes with

Positive Fraction	$\frac{\#\ \text{positive probe pairs}}{\text{total}\ \#\ \text{probe pairs}}$. Pairs used are equal to the number of probe pairs in the probe set minus the masked-out probe pairs. For a variety of reasons, the user may decide to mask out one or more probe pairs. Also, probe cells reporting intensities greater or less than 3 SD from the mean probe cell intensity in the probe set are masked out for being outliers.
Positive-to-Negative Ratio	$\frac{\text{Pos}}{\text{Neg}}$ ratio or $\frac{\#\ \text{positive probe pairs}}{\#\ \text{negative probe pairs}}$.
Log Average Ratio (Log Avg Ratio)	A number describing the hybridization performance of a probe set by determining the ratio of the PM to MM intensities for each probe pair, taking the logarithm of each of the resulting values, then averaging those across the probe set. This is a slight simplification as the reported Log Avg Ratio is also corrected to exclude the extremal outlier probe pairs in a probe set, *i.e.*, pairs with the largest and smallest contrast in intensity. It indicates random cross- or nonspecific hybridization with the higher values suggesting a higher likelihood that a transcript is detected.
Average Difference (Avg Diff)	The Avg Diff is a number calculated by taking the difference between the PM and MM of every probe pair and averaging the differences over the entire probe set. It corresponds to the absolute expression level of a transcript.
Absolute Call (Abs Call)	Has values A (Absent), M (Marginal) or P (Present) regarding the presence of a transcript and is determined from a decision matrix combining the Log Avg Ratio, $\frac{\text{Pos}}{\text{Neg}}$ Ratio and Positive Fraction.

Table 3.2
Affymetrix aggregate or derived measures per probe set

negative Avg Diffs are simply omitted from the analysis because the common log transformation used for robotically spotted array data is not defined for negative numbers. This loss of information is obviously not desirable. Other approaches such as thresholding negative numbers to a positive constant create a systematic artifactual bias in the distribution of Avg Diff results in the data set that affects all subsequent analyses that depend on these quantities.

Regarding the Abs Call per probe set, Affymetrix has empirically developed a decision table for each probe set of 16 to 20 probe pairs that is used to determine whether there was a "Absent, "Marginal," or "Present" call based upon the Positive-to-Negative Ratio, Positive Fraction, and Log Average Ratio. Note that this decision table will have different values for these three parameters for each chip set. The need for such a decision table stems from the fact that the specificity of

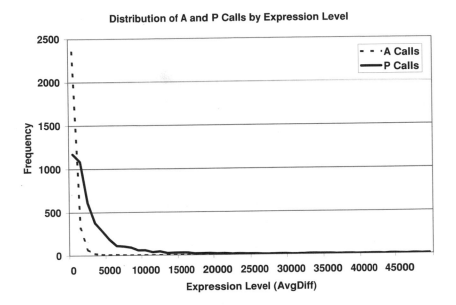

Figure 3.7
The Affymetrix *Absolute Call.* The relationship between Absolute Call and expression level for a microarray. Far more Absent (A) calls than Present (P) calls are at lower expression levels, but significant numbers of Absent calls are found even at expression levels in the low thousands.

the different probe pairs will vary across probes and across samples. In theory, the Abs Call provides a qualitative measure of reliability of the reported expression level (see figure 3.7). The naming convention of the Abs Call values is unfortunate as the report of an Absent Abs Call does not necessarily mean that the gene is not expressed in a sample; rather it signifies that the gene expression measurement on that microarray and for that experiment is not reliable, *e.g.*, its signal was below background effects.

A general understanding of these aggregate measures will make the reader aware of some major implications for analysis. First, the aggregate probe set values may not be reliable for all purposes. For example, in contrasting reported measurements from separate arrays, the aggregate measures Log Avg Ratio and Avg Diff may not be directly comparable *as is* because of the differences in specificity and sensitivity of each probe set, which is not captured by the aggregate measures. In contrast, Affymetrix software exploits probe-specific knowledge within the probe set in developing the decision table so that analyses and computations occur at the level of

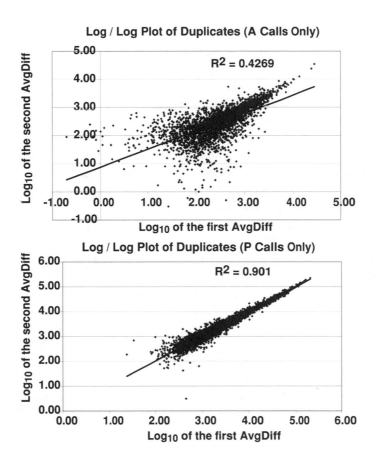

Figure 3.8
Graphical display of how Affymetrix Absolute Calls predict reproducibility. Correlation of expression levels from the same hybridization "cocktail" measured on two microarrays. Top: the correlation using only those genes for which an Absent call was reported by both microarrays. Bottom: only the Present called genes.

the `.cel` file. As described above, these files report the intensity of the individual probe cells, though at the time of writing, the oligonucleotide sequences for each probe cell are not known. Since most microarray analyses reported in the literature mainly employ Avg Diff values and not individual probe cell intensities, the comparability and reproducibility of these published results are less than optimal.

Another factor complicating the use of these aggregate measure values is that several settings, such as the noise threshold, are tunable by the operator of the Affymetrix scanner and software unit such that different reported aggregate measures may not be comparable across different microarray data sets. Despite the preceding concerns regarding aggregate measure values, the standard in the literature has typically been to not publish the `.cel` files with individual probe measurements but instead to release the aggregate statistic, *e.g.*, `.chp` files with all their attendant problems with comparability evolving from these.

At present and in sharp contrast to robotically spotted microarrays, the use of oligochips runs counter to the do-it-yourself ethos of many biologists for three primary reasons. First, except for large pharmaceutical companies, the construction of oligonucleotide microarrays occurs offsite and completely under the control and supervision of the manufacturers (essentially not peer-reviewed). Second, current oligochip manufacturers charge what the market will bear, and oligochips are typically more expensive than robotically spotted arrays. Third, the specific oligonucleotide sequences that are used by the manufacturers of oligonucleotides to interrogate an RNA transcript are proprietary information. It is likely that as the market for microarrays broadens, the second and third problems will resolve themselves. Regarding the first reason, in this genomic era in which we see the rapid industrialization of biological investigations, an ever-increasing fraction of the biologist's tools will become commodities available from vendors at diverse price-performance trade-offs. From that perspective, industrial fabrication of whole transcriptome microarrays will likely become common in the near future (see section 7.1).

3.2   Replicate Experiments, Reproducibility, and Noise

A ubiquitous and underappreciated problem in microarray analysis is the incidence of microarrays reporting nonequivalent levels of an mRNA or the expression of a gene for a system under *replicate* experimental conditions. This phenomenon of microarray data ir*reproducibility* is widely attributed to *noise* in the bioinformatics literature. For example, a common sample probe or target pool that is split and hybridized simultaneously to two separate chips of the same make and following the same scanning protocol will almost surely result in two, not exactly similar-looking, graphic representations of RNA abundance. These images will subsequently translate into two quantitatively different expression data files (*e.g.*, .cel or .chp files on the Affymetrix platform) after they are fed into image-processing and statistical software. This situation is akin to saying that two Northern blot assays for a particular RNA species in a split total RNA sample, having been processed under an identical protocol and conditions, led to different X-ray film images of the RNA intensity, thereby resulting in two different reports of expression intensity by a phosphor imager, which is clearly undesirable.

The ir*reproducibility* of a measurement reading—that is not entirely attributable to poor experimental design or practice—is neither new nor endemic to the microarray technology. One encounters this same problem in any measurement of a physical quantity, especially when these quantities exist on microscopic scales or when the measurement device is highly sensitive. Let us consider a thought experiment illustrating the latter point: Get 500 ml of well-mixed NaCl solution (*i.e.*, salt water). Split this solution into two samples A and B, and measure the NaCl concentration in both samples with a series of k measuring devices of increasing sensitivity which report concentrations up to an accuracy of 10^{-k}kg mL^{-1} for $k \in \{1, 2, \cdots\}$ respectively. Let $C_A(k)$ denote the NaCl concentration in sample A as reported by device k, and likewise define $C_B(k)$. When the device sensitivity is low, say, when $k = 0, 1, 2$, the device would probably not detect any concentration differences between A and B, *i.e.*, $C_A(k) = C_B(k)$ or more concretely, $C_A(0) = C_B(0) = 5$; $C_A(2) = C_B(2) = 5.01$, say. As the device sensitivity increases, say, for $k > 9$, typically negligible physical factors such as small pockets of uneven NaCl concentration in A and B might be within the device detection limit and be recorded so that $C_A(k) \neq C_B(k)$ or more concretely, $C_A(10) = 5.0100000001 \neq C_B(10) = 5.0100000002$, say.

The usual solution in these situations is to perform as many repeat or *replicate* measurements as are feasible and practical, with the hope that a statistic of these repeats converges asymptotically to the true measure of that quantity with the number of *replicate* measurements following the law of large numbers—provided certain conditions concerning the stochastic independence and distribution of these repeats are met. This approach may minimize, on average, certain systematic *noise*, such as measurement errors, but it cannot practically resolve more diverse manifestations of *noise* such as ones originating from biological variation in the model samples under investigation. Furthermore, there is the added issue of material cost with regard to replicate microarray measurements. Researchers today rarely perform more than 3 replicate chip assays per experimental condition due to the relatively high cost per microarray experiment.

In studies involving microarrays, *replicate* experiments are especially important for, but not limited to, these following reasons:

- Microarray data are often employed during the earliest stages of studies that are not hypothesis driven and are primarily exploratory in nature. In such cases, the attendant analyses and conclusions will typically provide hypotheses (*e.g.*, candidate ESTs) for more focused laboratory investigation at the later stages of the study. For these endeavors, further pursuit of a false-positive conclusion can potentially be costly in terms of resources and time.

- It is generally good scientific experimental practice, whenever possible and feasible, to replicate a experiment to verify an earlier, unconfirmed quantitation.

- Data from *replicate* experiments provide a better quantitative understanding of the extent of *noise* (see section 3.2.5) inherent in both the system under investigation and the measurement device [116]. The effect of *noise* is typically proportional to the level of sensitivity of the measuring device. As we have previously noted, the amount of total RNA for a Northern blot assay of a single RNA is more than enough for a typical microarray assay of more that 10^4 different mRNA species levels in today's technology. Furthermore, the *noise* effect might not scale linearly with the detected expression level and may be sequence dependent.

The terms *replicate*, *reproducibility*, and *noise* are explained in the next section.

3.2.1   What is a replicate experiment?   A reproducible experimental outcome?

> *No one can step into the same river twice.* Heraclitus, c. 540–480 B.C.

> *Nature does not make leaps.* G. W. Leibniz, 1646–1716

Consider the following five thought experiments that aim to compare the levels of blood glucose in 6-month-old mice, 1 hour before and 1 hour after an intravenous insulin treatment at noon.

E1. A 6-month-old mouse M has its glucose levels measured and recorded by a glucometer G, 1 hour before and 1 hour after an intravenous insulin treatment at noon. Call these readings $R(M, G, \text{Monday}, 11)$, $R(M, G, \text{Monday}, 13)$, respectively.

E2. A 6-month-old mouse M', not the same mouse as M, is subject to the same experimental protocol as for M in **E1** on the following day and here we call the readings $R(M', G, \text{Tuesday}, 11)$, $R(M', G, \text{Tuesday}, 13)$, respectively.

E3. The same 6-month-old mouse M in **E1** is subjected to the same experimental protocol the following day as it had undergone in **E1** with glucometer readings $R(M, G, \text{Tuesday}, 11)$ and $R(M, G, \text{Tuesday}, 13)$, respectively.

E4. Another 6-month-old mouse M', not the same mouse as M, is subject to the same experimental protocol as for M in **E1** simultaneously as M but using another glucometer G' to obtain readings $R(M', G', \text{Monday}, 11)$, $R(M', G', \text{Monday}, 13)$, respectively.

E5. During the time **E1** was carried out, M has its glucose levels simultaneously measured and recorded by another glucometer G' to produce readings $R(M, G', \text{Monday}, 11)$ and $R(M, G', \text{Monday}, 13)$.

Questions: Which of experiments **E2** through **E5** qualify as a *replicate* of **E1**? Which is "the best" *replicate* of **E1**? Observe that **E2** replicates the experimental protocol, except for the biological variation between mice M and M' and the difference in days. **E3** replicates **E1** in terms of experimental protocol and the use of the same mouse, except for the difference in days and perhaps more important, the initial condition (physiological) of mouse M on Tuesday could be measurably different following its Monday treatment. **E1** and **E4** are replicates up to experimental protocol and simultaneity in running both experiments, modulo the biological variation between M and M', and the use of separate glucometers which may be calibrated differently. Finally, **E1** and **E5** are simply repeat measurements of the glucose levels of M following the same protocol but with different glucometers. Typically in the study above, the absolute values $R(\text{mouse}, \text{glucometer}, \text{day}, 11)$ and $R(\text{mouse}, \text{glucometer}, \text{day}, 13)$ are themselves not important; rather, it is the relative change (difference or fold) from $R(\text{mouse}, \text{glucometer}, \text{day}, 11)$ to $R(\text{mouse}, \text{glucometer}, \text{day}, 13)$ that is of interest.

The point of the illustration above is to demonstrate that the question above is not well-posed until we specify what we mean by a *replicate*. Clearly, there are different levels or gradations for deciding how one experimental setup replicates another. *The way that one chooses to define a replicate experiment will be driven by, and context dependent upon the biological question that these experiments were designed to answer in the first place.* Only after one has made this definition can one sensibly discuss *(ir)reproducibility* of the attendant experimental outcomes. The range of parameters that could potentially enter into, be manifested, and detectable in a complex biological system is arguably wider and less easily characterized or controlled than in systems in the physical sciences. In our example above, the day of the experiment or the individual physical condition of the mouse may additionally affect its blood glucose level. Without entering into the epistemologies of Bacon [14] or Popper [143], the experimentalist normally attempts to resolve this difficulty by designing appropriate control conditions for the experiments and hope that – to paraphrase Leibniz's aphorism that measurable processes in nature do not change abruptly – the biological processes are not dramatically different from one normal mouse to another. Specific to our thought experiments, if the experimentalist has decided to replicate **E1** in the sense of **E4**, then she or he would reasonably expect the mice physiologies (and by extension, their blood glucose level in reaction to insulin) in **E1** and **E4** to not be radically different so that relative changes such as

$$\Delta(M', G') \;\doteq\; R(M', G', \text{Monday}, 13) - R(M', G', \text{Monday}, 11),$$

$$\Delta(M, G) \;\doteq\; R(M, G, \text{Monday}, 13) - R(M, G, \text{Monday}, 11),$$

are "close" to one another. A measure of similarity such as a *metric* is used to determine and quantify the "closeness" of $\Delta(M', G')$ to $\Delta(M, G)$, and thereby the *reproducibility* of the experimental results. Detailed discussions of measures of similarity are found in section 3.6. While it is highly unlikely in a real laboratory situation to find $\Delta(M', G') = \Delta(M, G)$, the experimentalist might expect that with a weaker measure of similarity—such as $\Delta(M', G')$ "=" $\Delta(M, G)$ when both $\Delta(M', G')$ and $\Delta(M, G)$ have the same sign—the experimental results are *reproducible*. Additionally, a statistical approach may be incorporated into the idea of reproducibility. For instance, if we carried out 1000 experiments **E1** with a different mouse each time and followed the same protocol, we may find that 990/1000 or 99% of the mice that were sampled exhibited a decrease in glucometer readings after treatment so that we can reasonably state that *most* – assuming a fair and random enough sampling – 6-month-old mice experience a drop in glucometer reading after an insulin intake at noon. Postulating (and *checking*) further properties about the data readout distribution, standard (non-)parametric statistical tools such as the Student t and Kruskal-Wallis tests may be used to determine the statistical significance of this conclusion, specifically the average drop in blood glucose reading.

3.2.2 Reproducibility across repeated microarray experiments: Absolute expression level and fold difference

Akin to situation E5 above, suppose that a common total RNA sample is split and the expression levels of the different RNA transripts within each split sample are measured on a separate microarray of the same lot. For such situations, the current bioinformatics literature typically reports the Pearson's correlation coefficient of expression intensities of one chip reading versus the other as being close to $r = 1.0$, *i.e.*, the separate readings are highly correlated, and used this to conclude that the experimental assay is highly reproducible. We recall that in order to obtain any sort of meaningful statistic out of the Pearson's correlation coefficient, one typically has to assume that the paired expression data point for each gene is independent of those of the other genes. But how are we to account for individual genes whose expressions as reported by each chip are not exactly equal?

To date, few studies focus on the reproducibility of microarray measurements [116]. In a publicly available document [9], Incyte Pharmeceuticals demonstrated a high concordance between RNA expression measurements on cDNA chips using Cy3 and Cy5 dye signals. Based on this finding, Incyte estimates the limit of detection of fold differences at 1.8, meaning to them that 95% of fold differences between samples of 2.0-fold and higher are significant From measuring the expression levels of 120 genes in various cancer cell lines using cDNA spotted filters, Bertucci *et al.* [24] showed that close to 98% of the measurements showed less than a 2.0-fold difference upon repeat. Richmond *et al.* [152], in their study of differentially expressed genes in *Escherichia coli*, filtered out genes below a minimum absolute expression threshold and ones with less than a 5.0-fold difference. Geiss *et al.* [75] used a Cy3/Cy5 system to measure genes differentially expressed during human immunodeficiency virus (HIV) infection in which they determined that fold differences of as little as 1.5 were statistically significant. This minimal fold threshold was determined by excluding 95% of the expression measurements that were reported, and not by using information-theoretic methodologies. Publications citing fold differences between control and test groups which are as low as 1.7-fold continue to be published, *e.g.*, [117].

The foregoing comparative studies do not address the reproducibility of fold differences if *entire* microarray experiments were repeated. Butte *et al.* [40] investigated this topic using Affymetrix Hu35K oligonucleotide microarrays containing 35,714 unique probes to measure RNA expression levels of muscle biopsies from 4 patients: P_1, P_2, P_3, P_4. Duplicated measurements were made from each patient sample: P_1', P_2', P_3', P_4'.

A linear regression normalization procedure—outlined in section 3.4.1—was applied on all the chip data with respect to the data from patient P_1. *Intra*-patient logarithmic fold differences (LFDs) were then calculated between the duplicated measurements for each of the four patients. Furthermore, *inter*-patient LFDs were calculated between all 6 possible pairs of patients (P_1 vs. P_2, P_1 vs. P_3, *etc.*) and their duplicates measurements (P_1' vs. P_2', P_1' vs. P_3', *etc.*). The LFD was used throughout this analysis so that the levels of up and down fold regulation were

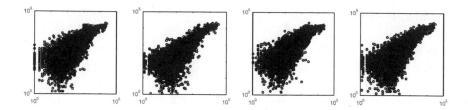

Figure 3.9
Expression measurements made in duplicate from the same RNA samples do not correlate well all the time. RNA samples from four human samples were placed on duplicate oligochips and the expression of 35,714 ESTs was measured. Each point represents an EST. The duplicate expression measurements are plotted here on a log-log scale (base 10). $r = .69, .73, .73, .69$. (From Butte *et al.* [40].)

numerically equal in magnitude but oppositely signed.

Ideally, if the oligochips assays were perfectly reproducible across the duplicates, one would expect that Pearson's correlation coefficients at 1.0 between the four repeated expression measurements, and intra-patient LFDs calculated on each of the duplicated measurements to be the same, *i.e.*, if gene A is noted to be 10 times higher in P_2 versus P_1, then it should ideally and similarly be 10 times higher in P_2' versus P_1'. However, the Pearson's correlation coefficients between duplicate expression measurements of the four patients across all 35,714 probes were .76, .84, .78 and .82 (see figure 3.9). Furthermore, the cross-probe Pearson's correlation coefficients for the six paired (replicate) inter-patient LFDs were near .0 as shown in figure 3.10 where the first graph shows the plot of $\log(P2/P1)$ versus $\log(P2'/P1')$. Further analysis showed that this poor correlation coefficient in the replicated LFD was primarily due to small absolute expression values: When folds are calculated using a pair of numbers where the denominator quantity is small (*i.e.*, the gene was found by the microarray to be not highly expressed), a high fold difference is typically the result. This is particularly problematic as the effect of noise in microarray assays today appears to be more pronounced at lower absolute expression levels.

It is worth noting that even when the reported absolute expression measurements are reasonably reproducible between replicate measurements, the fold differences calculated from these reported measurements may not be as reproducible across replicate experiments. As mentioned above, the fold or ratio of these expression quantities is very sensitive to small numerical perturbations, especially in its denominator quantity. This is best explained with an illustration: Say that we carry out a pair of experiments P_1, P_2, and that the expression levels of 2 genes G_j and G_k re-

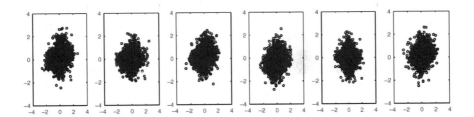

Figure 3.10
When expression measurements do not correlate well, fold differences correlate even poorer. Fold differences of 35,714 ESTs were calculated between the six possible pairings of the four patients. Fold differences are expressed in base 10 logarithm, so that ESTs that did not change between models are plotted in the center of each graph. Fold differences from the duplicated measures are shown on the x- and y-axes. Even though the correlation coefficients were high between original and repeated expression values, the correlation coefficients were very low between original and repeated calculated fold differences. (From Butte *et al.*, [40].)

ported in each experiment are $(1.0, 2.0)$ for G_k and $(500.0, 1000.0)$ for G_j, so that the fold change for both genes from P_1 to P_2 are both $2.0 (= 2.0/1.0 = 1000.0/500.0)$. Suppose we further carry out of pair of replicate experiments P_1', P_2', and that the reported expression levels are now $(1.5, 2.0)$ for G_j and $(500.5, 1000.0)$ for G_k. Note that the denominator quantity in the replicate P_1' experiment is perturbed by $+0.5$ from the reported measurement in P_1—an effect one typically sees in the presence of noise. Let us suppose that the overall absolute expression measurements P_1 to P_1' and P_2 to P_2' are reasonably reproducible, *i.e.*, the intensity-intensity Pearson's correlation coefficients between the duplicate experiments are greater than .99. Note, however, that the resulting fold changes calculated using these new values are now $1.333 (= 2.0/1.5)$ for G_j and $1.998 (= 1000/500.5)$ for G_k. Clearly, the robustness of the fold is dependent upon the absolute expression measurement, as we see here that the fold of gene G_j is more stable to noise than gene G_k. For a detailed discussion of the appropriateness of using the fold as a measure of expression change apropos microarray data, we refer the reader to section 3.5.

The key points of this subsection are as follows:

- When entire microarray experiments are not replicated, one would not be aware of, or be able to quantify, the extant irreproducibility in the experimental data and design, and false biological conclusions may potentially be generated.

- Even if the gene expression measurements from a single sample placed on two microarrays appears to have reasonable correlation coefficient, the fold differences calculated using those reported measurements may not reproduce as well between replicates at [40].

Based on this experience, we now recommend at least two to three replicates for each data point. Triplicates are certainly preferable for in worst-case scenarios, it is unclear which of the duplicate measurements is more correct. Preferably, the RNA sample for each replicate should be generated separately to maximize noise inclusiveness in the replicate model.

3.2.3 Cross-platform (technology) reproducibility

The process of comparing gene expression information from a broad range of conditions has been widely adopted as an approach to functional discovery. As each individual laboratory may use different expression assaying platforms to profile a common subset of genes, one obvious question arises as to how one may combine and compare expression measurements across diverse technologies. Assuming that such operations makes sense, the problem comes down to determining the reproducibility of cross-platform measurements. Analyses by Kuo [113] suggest that for spotted cDNA probe microarrays and microarrays with Affymetrix-synthesized oligonucleotides circa 1999, the reproducibility may be poor. The mRNA expression measurements from oligonucleotide microarrays from Butte *et al.* [39] were paired with measurements from the cDNA microarrays from Ross *et al.* [154] by matching the corresponding cell lines and RNA transcript probes. When the oligonucleotide data (from the Avg Diff column) were compared to the normalized Cy5/Cy3 ratio (from the RAT2N column in the ScanAlyze output file) of the cDNA data, the overall linear correlation across all 162,120 paired measurements was poor ($r = .0326$) as illustrated in figure 3.11. Corrections for GC content, cross-hybridization, and probe length all did not materially improve the correlation coefficient.

It does not seem likely that the reported measurements from these data sets can be normalized or transformed into a single standardized unit of gene expression. Moreover, even the qualitative correlation was poor. For instance, high measurements in the oligochip did not correspond well with high measurements in the cDNA chip, and *vice versa*. At the very least, the results of this study suggest that greater (pre)caution should be taken in making cross-platform comparisons. While complementing existing assay technologies, new techniques for expression profiling such as optically coded beads and ink-jet arrays (see section 7.1) will add to the challenge of comparing and combining expression data across different platforms, and are likely to require similar efforts for normalization and standardization.

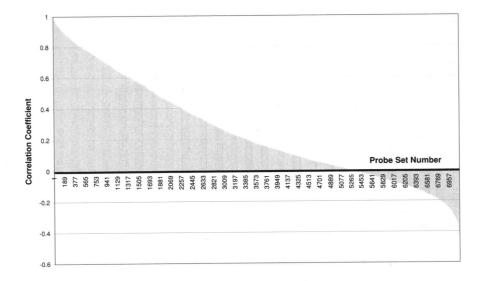

Figure 3.11
Overall correlation coefficients for each matching probe and probe set across two microarray technologies. Great variance is seen in reproducibility of measurements across probes and probe sets, with some probes showing correlation coefficients near 1.0, and some even showing a negative correlation (*i.e.*, a gene which is reported as highly expressed in one technology has an opposite expression report in the other technology).

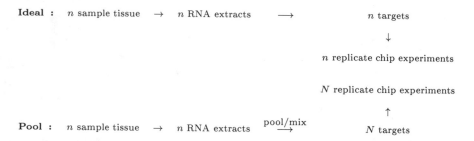

Figure 3.12
Pooling total RNA extracts for replicate experiments. Two strategies for making replicated measurements from samples, pooled and not pooled. N may not necessarily equal n.

3.2.4 Pooling sample probes and PCR for replicate experiments

A typical microarray hybridization experiment today requires on the order of 10 μg of total RNA to generate the labeled sample probe compound for each array (spotted cDNA array, see [167]; oligochip, see [122]). Depending upon the experimental design, namely, how one defines a replicate experiment, it might not always be possible to obtain enough total RNA from a system, *e.g.*, specific tissue from a test animal, to prepare sample probes or targets for one or more (as repeat measurements) chip experiments. For example, suppose that a biologist wishes to obtain the gene expression profile of the mouse heart during embryonic day 12 (E12) and finds that the total RNA extracted from one embryonic mouse heart is not sufficient for generating enough labeled target for even a single microarray experiment. In this situation, an alternative solution is to pool together the total RNA extract from several, say n many, different E12 mouse hearts to create a sample probe or target compound which is then split into N ($\in \mathbb{N}$), $N < n$, many targets for hybridizing on N separate chips, and hence resulting in N replicate experimental data.

When these tissue samples are drawn from n embryos that do not share a common parent, the pooling procedure will average out *intra*-species biological variations between the mice. This is independent of and in contrast to the subsequent splitting of one pooled sample into individual targets for N separate microarray assays and takes into consideration measurement error that inevitably arises in any empirical experiment.

Regarding the use of PCR to amplify these and any RNA extracts, the reader should note that beyond a small number of amplification cycles, a heterogeneous RNA mixture that has been amplified via PCR may not retain the same specific RNA percentagewise composition as in the original mixture [17]. Furthermore, one should also note that experimental data generated by re-using a target that has undergone a previous hybridization should not generally be considered a replicate assay because the biochemical composition of the target may have changed irreversibly after a hybridization procedure, due to both the protocol and time differences. Typically, a target needs to be incubated on a microarray for 12 hours or more during hybridization.

3.2.5 What is noise?

Noise is an informal term widely used in microarray data analyses and experiments to refer to:

N1. Physical effects and parameters extraneous to the aspect of a system one is investigating, which have entered into the system or the measurement device that quantifies this aspect, and which, if detectable, would lead to,

N2. Experimental outcomes which contradict *a priori* known facts, or more specifically and commonly, variations in experimental outcomes that should be "identical."

Strictly, **N1** and **N2** are origin (or cause) and effect respectively, *i.e.*, **N1**, if detectable by the measurement device or microarray, is manifested to the laboratory observer as **N2**. Most studies do not distinguish the former from the latter in their use of the term *noise*. Typically, the presence of noise in a system is deduced from comparing an actual measurement outcome of the system against another comparable reference quantity or a known and confirmed fact. For instance, if we use a ruler to measure the length of a pencil twice and obtain slightly different readings each time, then we know that there must be "noise" in our measurement activities because the indisputable *a priori* fact is that we are measuring one and only one pencil, with a single fixed length. Here, **N1** might include parallax error and **N2** is the different recorded lengths. Recalling our prior discussion on replicate experiments, the range of parameters that affect a biological system study is vast and not completely controllable so that, in general, **N1** may encompass any combination of conditions that are present at the time of a microarray assay, whether they are inherent in the measurement device or in the biological system.

For partly cultural reasons, there is a lack of appreciation for the challenge presented by noise in microarray expression data analysis. Historically, genomicists were fortunate to have had their first and successful foray into understanding patterns in the genome to have occurred at the level of the DNA sequence. DNA sequence components, being drawn from an alphabet of 4 characters (A, T, C, G), are unambiguous—and furthermore, when one assumes that these sequences code for function deterministically, one finds that the representation of information in a DNA sequence is essentially *digital*, or discrete. By contrast, the measure of mRNA species levels, and by extension the expression levels of associated genes, is intrinsically *analog*. Because of this, there is more room for measurement error, and a greater and generally unfavorable dependency on the measurement technology. RNA level is a continuous quantity and its measurement is no different from the measurement of any other biochemical processes or products, such as the rate of lactic acid oxidation in muscle or endorphin levels in the bloodstream. Specifically, in the context of microarrays, each chip measures microquantities of upwards of 10^5 different chemical species at a time on a hybridization medium with an area of less than 10^{-4}m^2. Bioinformaticians who make the transition from sequential to expression genomics are often surprised by the massive increase in the ambiguity and noise levels in the genomic data.

It is worth noting that noise is not entirely a negative phenomenon. In some situations, it could inform the genomicist to reconsider *a priori* assumptions, postulates, or questions for a system which may not be entirely correct or well posed. For instance, referring to the earlier thought experiments for replicating the measurement of blood glucose levels in mice in section 3.2.1, while differences (noise) in the repeat measurements **E5** point to measurement errors which are unquestionably undesirable, differences in **E1** and **E4** (that are not measurement errors) might imply the existence of significant physiological variations between M and M' that may suggest that not all mouse glucose levels respond similarly to intravenous insulin and this knowledge inevitably refines or changes the study question and objective.

3.2.6   Sources and examples of noise in the generic microarray experiment

Before delving into details of how one can distinguish chip-reported expression data that arise from a genuine measurement of mRNA levels in the presence of noise from reported data that come from noise alone, it is informative to review the ways in which noise could enter a typical microarray experiment. Noise can roughly be classified, by point of origin, into two categories:

- *Intra*-**chip noise**   encompasses the previously discussed *background effect* on the chip surface, where localized physical and chemical changes in one probe feature diffuse into neighboring probe features. Improper techniques in scanning are also included, such as having one area of the chip brighter than another. Chip manufacturing defects such as quantitative nonuniformity in laying probe sequences on the array surface would also be covered (see figure 3.18).

- **$Inter$-chip noise** includes biological variation in the samples (*e.g.*, note that even pure cell populations are unsynchronized in their cell cycle phase). Also included are protocol variations in the hybridization procedure, the environment, time effects, and again, possibly, inconsistencies in chip manufacture.

In contrast to $inter$-chip noise, $intra$-chip noise is often more subtle and not as easily discernible, due to the fact that there is usually no obvious reference situation for a comparison of a single microarray experiment outcome.

Next, we explore examples of noise in microarray data. Figure 3.13 below, courtesy of Todd Golub at the MIT/Whitehead Institute's Genome Center, displays the gene expression intensity-intensity plot from an experiment that has been repeated with the same sample hybridized onto two separate arrays of the same type. First, it should be pointed out that had the two arrays produced exactly the same data, a plot of the gene-by-gene expression level of chip 1 versus chip 2 would have shown the data points aligning themselves exactly along the diagonal line $y = x$. Instead, here we see a typical, and, in fact a rather qualitatively good, result for intermicroarray variation. This variation is a very typical manifestation of $inter$-chip noise. Note especially that the typical data point spread away from this $y = x$ line increases with decreasing expression intensity. As mentioned earlier, this is due to fact that at lower gene expression levels, the corresponding RNA amounts that are being measured are smaller and therefore the effect of noise in these measurements is relatively more significant. Thus, there is greater variation here between these two nominally identical probe assays in a repeated assay.

Golub *et al.* [78] also plotted two lines, $y = 2x$ and $y = \frac{1}{2}x$, which mark the locations of a 2.0-fold increase and a 2.0-fold decrease in gene expression, respectively, relative to the $y = x$ line of an idealized 1.0-fold increase (or decrease), *i.e.*, zero variation. Observe that at lower expression intensities, there is a sizeable number of genes whose expression varies outside of this 2.0-fold envelope despite the fact that these intensities came from essentially identical samples. A similar-looking pattern is seen in the following two figures. The second figure shows data from repeat hybridization of a sample that was hybridized to one chip, washed off from the first chip, and then rehybridized onto a second chip. In the third figure, RNA was extracted twice from a common sample and hybridized onto two different chips from the same lot simultaneously.

In all three scenarios, there is a similar and more prominent spread of data points about the $y = x$ line at lower intensity levels about the idealized line of zero variation

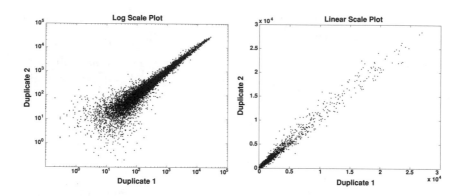

Figure 3.13
Apparent differences in source of variance due to log scaling. On the left is the apparently
increased variation at lower expression levels using a plot of the logarithm of all the gene
expression levels measured in one microarray versus the logarithm of all the of the genes
measured on another microarray. When the expression values are plotted on a linear scale as on
the right, then it appears that the higher expression levels have higher variance. The former
scale emphasizes the decreased reproducibility at lower expression levels and the latter
emphasizes that there are fewer measurements or genes at higher expression levels. Note that for
well-definedness, the logarithmic scale plot excludes measurements that had a negative value.

$y = x$. At this point, we bring to attention a common *trompe l'oeil* that derives from
the logarithmic scaling commonly employed to illustrate the concordance (or lack
thereof) between two microarray hybridization data sets. As shown in figure 3.13,
by scaling the data sets to a logarithmic scale, the apparent variance at the higher
expression levels is reduced. However, as we will explore further in our analysis of
noise in microarrays, it remains that a larger proportion of the poorly reproduced
results with microarrays are obtained at the lower expression levels. The reason for
the apparent wider scatter at higher expression levels in the non-logarithm-plotted
graph in figure 3.13 is that there are relatively fewer genes with high measured
expression levels.

In figure 3.8, we illustrate another example of the significance of noise in oligonucleotide microarrays, the Affymetrix GeneChip. Here, we show measured expression
data or Avg Diffs plotted from an experiment where the same RNA sample was
hybridized onto separate oligochips for human skeletal muscle. All gene express-
sions were plotted on one of the three graphs based upon their Affymetrix Abs
Calls (Absent, Marginal, Present). As we noted in section 3.1.2, Affymetrix Abs
Calls are determined from a decision matrix which is a function of the specificity
of probe set expression for each gene. Note that the decision rules for this matrix
are not currently available to the public. The lack of robustness or reproducible

GEMID	Location	DiffExpr	BalancedDiffExpr	P1Signal	P1S/B	P1Area%
022PAOWV	2196	1.2	1.2	4330	60.3	90
022PAOWV	783	-1.1	-1.2	4325	43.8	94

P2BalancedSignal	P2Signal	P2S/B	P2Area%	Probe1	P1Description	Probe2
3724	3616	70.5	90	12323996	cocsxplusminus.si	12363997
5001	4856	64.1	94	12323996	cocsxplusminus.si	12363997

P2Description	GeneID	PlateRow	PlateCol	PlateID	CloneID
csx	-137302	C	11	021OA0OE	524442
csx	-131764	E	5	021VAOMR	463651

CloneSource	AccessionNum	Locus	IncyteCloneID	PCRStatus	GeneName
IMAGEConsortium	AI325648.1	02:15.5	mm45a10	Passed	RibosomalproteinL7{IMAGE:524442}
IMAGEConsortium	AA027730.1		mi15h10	Passed	PublicdomainEST{IMAGE:463651}

Figure 3.14
Incyte data file snippet. Heads and the two rows of values from an Incyte data file. Incyte microarrays are constructed using a robotic spotting process. The spotting process involves two proximal spots for the two dyes—Cy3 and Cy5—one of which is the targeted clone and the other a control spot.

stability in both the Abs Calls and the Avg Diffs is evident from these plots. That is, for the same RNA sample both the expression level and the Abs Call will vary. Apropos the Abs Call instability, note that the total number of genes across the three experiments is well under the total number probes for each GeneChip. While such observed variances are not unexpected, it should warn the investigator against depending too heavily upon a particular microarray-reported value, statistic, or call in his or her research program.

The noise example above is not singularly unexpected or unusual within the gamut of microarray noise examples, nor is it particular to the Affymetrix technology. Similar manifestations of noise, though from possibly different sources, are just as consequential in other microarray technologies. For instance, consider data from robotically spotted Incyte GEM microarrays in figure 3.14, which are three rows of expression data from an Incyte file. Of relevance to our discussion are the location index of a probe on the physical microarray and its corresponding difference in expression. Incyte GEM microarrays are used in a technique similar to the two-dye fluorescence technique for cDNA chips pioneered at Stanford University, so that each microarray is hybridized to two differently labeled sample probes.

Since cDNA microarrays, unlike oligochips, allow for two sample probes corresponding to different experiments to be hybridized simultaneously on the chip, the *inter*-experiment noise here may also *intra*-chip rather than exclusively *inter*-chip as we had seen with oligonucleotide microarrays. The two different colors or labels correspond to signals P1S and P2S respectively. These signals are corrected for background intensity and its variation, B, by the image analysis software used with the Incyte microarrays. Background corrected values are reported as P1S/B and P2S/B which correspond to the background-corrected intensities for the two probes. The P1S and the P2S area percentages correspond to the areas of potential hybridization that are recorded by the image analysis software to have undergone actual hybridization. In theory, if the appropriate control probe is chosen, say P2S, then this should correct for changes in hybridization conditions across different areas of the microarray that is caused by noise, such as localized temperature differences on the microarray surface.

The first thing to note from figure 3.15 is that we see an almost similar distribution of intensities for an identical RNA sample (the "hybridization cocktail") that is hybridized onto two separate Incyte arrays, P and P'. One way to view the noise profile of a robotically spotted microarray is as a plot of the ratio of signal P2S versus P2S/P2S' as in figure 3.16. In the ideal situation, the latter ratio ought to be uniformly equal to 1.0. That is, data points should only be distributed along the horizontal line $y = 1$ for any signal intensity P2S. In fact, at lower signal strengths of P2S, there are a number of genes whose P2S/P2S' ratios are significantly above 2.0-fold, even though on average the number of these outliers decreases with increasing P2S, and most genes tend to have a fold of 1.0. This demonstrates once again that the expression variance at lower expression values is more pronounced.

Microarray-reported data may also vary by the spatial location of the probe on the physical array. In order to see this, we map every probe and its associated reported expression amount from the two dimensional physical chip onto a one-dimensional array or line, and plot the expression level of each cell on the y-axis versus its location index on the x-axis. Figure 3.17 displays such a plot of experiments involving Incyte microarrays. A visually prominent feature in these graphs are four spikes marking dramatic increases in the expression level of the four probe cell locations on every chip. Furthermore, there appears to be a periodic pattern in reported expression level with respect to probe location on these microarrays. A careful review of chip manufacturer's specifications reveals that there have indeed been four control probe sectors (spiked controls of reference RNA transcripts) on each chip which would account for these spikes. The first preanalysis task then is to remove these spiked controls' data from their respective data sets and to renormalize these data sets for background effects, thus producing the next plots. In figure 3.17 the fold differences appear to be much less variable than previously and less obviously influenced by the probe location on the microarray .

An eyeball inspection across thousands of data points as we have done above with the Incyte expression data set is often not sufficient for detecting subtler and

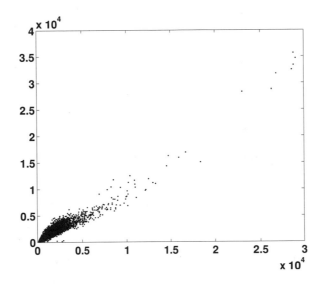

Figure 3.15
Duplicate cDNA assays of a common extract. The same RNA extract or hybridization
"cocktail" measured on two Incyte spotted arrays. The P2S signals (the second of two dyes)
from each microarray is plotted for each gene.

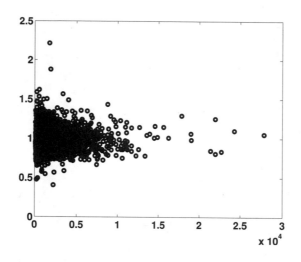

Figure 3.16
Ratio of probe intensities as a function of expression level. The ratio of the P2S/P2S' signals
(y-axis) from two robotically spotted microarrays plotted against the P2S signal. The same
RNA extract was used in both microarrays. Although the ratio is close to 1.0 for many genes, it
does deviate sporadically and widely in a few instances. Also, as the expression level decreases,
the distribution of the ratio spreads increasingly away from 1.0.

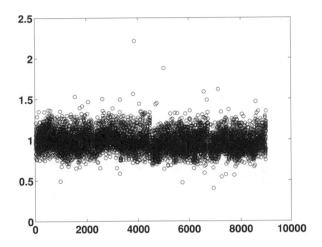

Figure 3.17
Ratio of expression level as a function of position. The ratio of the two *different* dye intensities per gene for a single microarray plotted against the position in the array (defined as a single number that is computed from the x and y coordinates of the probe on the array). The "control islands" that contain "spiked" controls on the array were not included in this plot. Also, all expression values were corrected for background noise. No particular pattern is obvious on inspection.

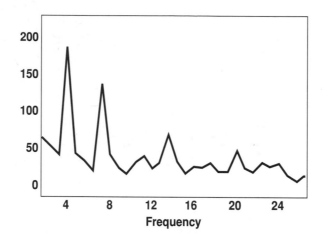

Figure 3.18
Fourier analysis of spatial series on spotted microarray. The same data illustrated in figure 3.17 is subjected to a Fourier analysis to extract systematic periodic signals. As shown, the frequency with the largest power is 4. That is, there appears to be a periodicity of 4 per microarray based on the position of the probe on the microarray. The magnitude of this systematic variation dominates fold changes in gene expression of the magnitude reported elsewhere as noteworthy.

less visually prominent patterns in the data representation. By applying standard signal-processing techniques such as the discrete Fourier transform, we find several major frequency peaks in the expression level across the chip (see figure 3.18). The highest peak here occurs at a frequency of 4 per array surface. This kind of periodic pattern in the Fourier transform expression data appears to be location dependent, and should lead the analyst to suspect that something systematically amiss occurred during the generation of this data set which could have happened during the manufacturing, hybridization, or the scanning stages.

Consider the following image of the Incyte GEM microarray at the time this analysis was done (figure 3.19). The pattern seems to suggest that the periodicity in the reported data originated from four separate but similarly behaved sources. It turns out in this example that during the array manufacturing process, four separate pins had indeed been used to spot each microarray quadrant. It should not be surprising, then, that small but systematic physical and chemical variations in the amount of probe cDNAs that were picked up and deposited by each of the four pins are reflected and recorded in the overall expression pattern one sees here. Other possible sources of the lower-level systematic variations seen in figure 3.18 may include the frequency with which the pin assembly units are replaced during

Figure 3.19
Incyte GEM Microarray *circa* 1999. In this photograph of a GEM microarray, note that there are four quadrants which are each spotted with separate pins.

the microarray spotting procedure.

Again, although these noise examples were taken from Incyte GEM microarrays and are of the *intra*-chip variety, it demonstrates that data from any microarray assay should be subject to at least some scrutiny or a quality control step as we have done here to screen out inherent and systematic biases in the measurement system. Numerous manifestations and sources of data acquisition error can be immediately caught by a simple visual inspection of the scanned images of a microarray. The images of the hybridizations illustrated for three different kinds of microarrays in figure 3.20 should be highly motivating in this regard.

3.2.7 Biological variation as noise: The Human Genome Project and irreproducibility of expression measurements

As of 2001, preliminary results of the Human Genome Project have provided further explanations and reasons for microarray measurements having poor reproducibility in addition to the ones iterated in section 3.2.5. We now know that there exists a relatively high frequency of *alternative splicing* products from each gene—see fig-

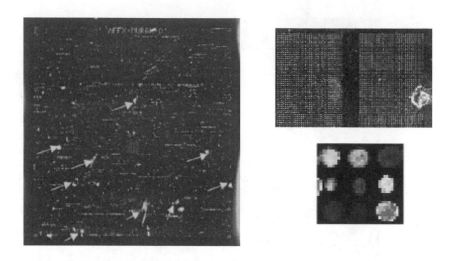

Figure 3.20
Problems in images acquired from microarrays. On the left: a contaminated D array from the Murine 6500 Affymetrix GeneChip set. Several particles are highlighted by arrows and are thought to be torn pieces of the chip cartridge septum, potentially resulting from repeatedly pipetting the target into the array [156]. On the right top: local changes in intensity due to contaminants and scratches. (Derived from http://www.mediacy.com/arraypro.htm.) Right bottom: high magnification of a scanned image of a spotted microarray. Note the different sizes and shapes of the spots [199].

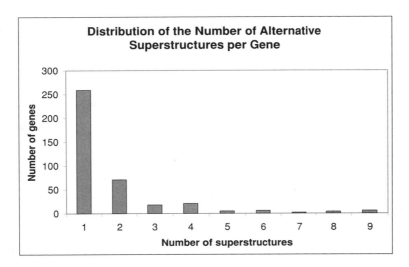

Figure 3.21
Estimate of the incidence of alternative splicing in genes. While most genes have only one
spliced product, many have more than one, and this can affect the the gene product detection
efficacy and rate by microarrays. (Derived from Mironov *et al.* [132].)

ure 3.21—and are more aware of the heterogeneity of common gene polymorphisms
across individuals and populations.

Genes, which are composed of exons and introns, have the remarkably versatile
property that their exons can be selectively added or deleted during transcription,
giving rise to different proteins from the same gene. This form of *alternative splicing*
appears to play a crucial role in the cellular physiology of higher organisms. For
example, in the development of *Drosophila melanogaster*, a single gene determines
the eventual sex (male or female) of the organism, depending upon how the gene
is spliced [91]. In contrast to the most simple eukaryotes such as yeast, it appears
that higher organisms such as humans have a much larger number of alternative
splicing products per gene. The average number of distinct transcripts per gene
averages 3.0 for humans as compared to 1.1 for yeast. Current oligonucleotide
microarrays only target a set of subsequences for a particular individual transcript,
and full-length cDNA microarrays only target a particular individual transcript.
Consequently, both these technologies will show variation in signal intensity even
if the same gene is being expressed at the same level but an alternatively spliced
transcript is being transcribed rather than the splicing for which the microarray
was engineered. Future microarrays may need to be constructed in a manner that
is exon- and intron-specific, instead of gene-specific.

Apropos the heterogeneity of gene polymorphisms, there is, on average, one single nucleotide polymorphism (SNP) in every 1000 human nucleotide bases. Polymorphic variations of the same gene can be quite common, as noted by Chakravarti [43], and as a result, a cDNA or oligonucleotide probe that is engineered for a particular variant might hybridize considerably less effectively with another variant. This could be true even if the polymorphism does not change the final protein product of a gene. Although the genetic code is somewhat redundant in the sense that 64 codons code for 20 amino acids,[4] two codons that code for the same amino acid may bind significantly differently to their complementary DNA molecule and these chemical differences might be detectable by the microarray.

3.2.8 Managing noise

Now that we have established that the current microarray technologies are bedeviled with *noise*, we will begin to address the data that are generated using such systems. The most direct route would be to have a direct measurement of the variation per gene of all microarray measurements. This is done by replicating the microarray assay(s). In practice, only duplicate or triplicate hybridizations are performed because of cost. The goal here is to obtain a subset of genes from these experiments whose reported expression level is robust across replicates, despite the noise. As we have alluded to in section 2.9 and as illustrated in figure 2.9, any of these techniques will inevitably have to balance between their significant false-negative or false-positive rate. That is, in order to avoid having to follow thousands of red herrings, as would be the case with a high false-positive rate, one has to be willing to discard many genes that might be genuinely involved in the biological process of interest on the grounds that their reported expression is inconsistent or irreproducible across replicates in the current assaying technology. We describe in this section, as an example, one technique that was developed to identify sets of genes that have been up- or downregulated, that had a low false-positive rate for the specific data set in consideration. When we consider the broader question of what constitutes a significant change in expression level for a gene, then the question of appropriate (dis)similarity measures is raised. This question is addressed in sections 3.3 and 3.6.

Modeling the expression-level-dependent noise envelope If, as has been mentioned in preceding chapters on reproducibility and replicates, the variance in reported gene expression level is a function of the expression level itself, then

---

[4]Twenty-one amino acids if one includes selenocysteine.

clearly single-value thresholds (*e.g.*, a 2.0- or 3.0-fold increase) applied to an entire expression data set, as is commonplace in the literature, are not sufficient for deciding which genes are significantly changed between different samples or conditions. At lower expression levels, these uniform thresholds will be associated with an increased false-positive rate, and at high expression levels with an increased false-negative rate. One way to address this problem would be to develop fold thresholds that are themselves a function of the reported expression level, rather than a fixed number. This is simply done by obtaining at least a duplicate hybridization from the same hybridization cocktail hybridized onto two microarrays. As much as is possible, these are identical experiments in which the operating conditions, cell lines, culture media, incubation time, and so forth are controlled to be the same.

As the expression values should be identical, all variance is theoretically attributable to nonbiological sources. Then, for each set of duplicates, an identity mask (ID mask) is calculated in Tsien *et al.* [181] wherein are fold changes which are insignificant or attributable to noise alone. Two parameters are used for creating each ID mask: expression value range (or a sliding window of expression intensities), and either the scale value or the number of standard deviations. These are used to calculate the ID mask borders and can be adjusted for different trade-offs in sensitivity and specificity, depending on one's utility model as reflected in figure 2.9.

We illustrate the application of ID masks to six experiments (A, B, C, D, E, F) that were run in duplicate. Total RNA was isolated from the cell lines (MCF-7 human breast cancer cells and MG-63 human osteosarcoma) and hybridized onto Atlas Human cDNA Expression Arrays from Clontech (Clontech Laboratories, Inc., Palo Alto, CA). Each of these Atlas Arrays (Human 1.2 I, Human Cancer) is a nylon membrane on which approximately 1200 human cDNAs have been immobilized. Although this example uses a relatively low-density array technology, the methodology applies similarly to higher-density microarrays.

Two methods are then explored for creating ID masks. Method 1 relies on segmental calculation of standard deviations. A "data point" refers to an (x, y) pairing, in which x is an expression value of a gene g from the first hybridization, and y is the corresponding fold difference value (*i.e.*, the ratio of the expression of the same gene g in the second hybridization to x). Using all data points in a given sliding window of expression values (*e.g.*, from intensities 1001 to 2000), the standard deviation of the fold values is calculated. The average of expression values within that intensity window is then paired with the average fold value within the same window plus the number of standard deviations specified by the experimenter. This new pair becomes a candidate "upper mask border" point. Similarly, a candidate "lower mask border" point is created by pairing the average expression value of that

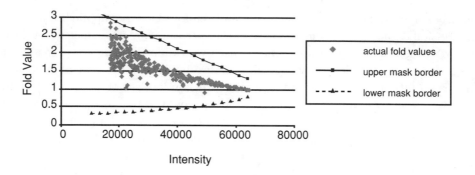

Figure 3.22
Identity mask for experiment A. Method 2 with parameters 9000 for expression value sliding window size and scaling factor 0.975 resulted in the lowest percentage of original data points lying outside of the mask region (0.7%).

window with the average fold value minus the number of standard deviations as specified by the user. Each successive group of data points in each sliding window of expression values (*e.g.*, all points from intensities 2001 to 3000, then all points from intensities 3001 to 4000, *etc.*) likewise give rise to candidate mask border points. A line is then fitted via least squares linear regression on the set of (expression value, fold value) pairs comprising the candidate upper mask border points. This line defines the upper mask border. Similarly, one computes the lower mask border from the set of calculated candidate lower mask border points. If one of the derived mask borders fits poorly—based upon its relationship to the original data points— the "reciprocal reflection" of the other (good-fitting) mask border can serve in its place. This simply means that each (x, y) point on the good-fitting (linear) border gives rise to a point $(x, 1/y)$ to create the reciprocal reflection border. Figures 3.22 and 3.23 show ID masks delimited by one linear regression border; the other border was derived by taking the reciprocal values of that linear regression border. The region between these borders represents the "identity" region of insignificant fold differences (*i.e.*, fold changes resulting from noise alone).

The second method, method 2, for creating an ID mask uses candidate mask border points derived from maximal points in each sliding expression intensity window rather than from standard deviation calculations as in method 1. Specifically, among all data points in a given intensity window, the point with the greatest fold value is chosen. This is repeated for each successive window of expression values. These fold values can also be scaled before use in a linear regression to find the upper mask border. The lower mask border is analogously derived from the min-

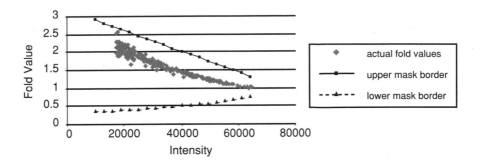

Figure 3.23
Identity mask for experiment E. Method 1 with parameters 5000 for intensity window size and 3 SD resulted in the lowest percentage of original data points lying outside of the mask region (0.9%).

imal fold values. Once the ID mask has been derived, all original data points are checked for inclusion or exclusion in the ID mask region. The percentage of data points lying outside of the ID mask region is recorded. We then automatically searched a large number of masks for those which provided the best performance. For both methods 1 and 2 of ID mask creation, sliding windows of ranges 1000, 5000, and 9000 on the expression value axis were chosen for experimentation. Only when calculations were not possible with one of these window sizes (*e.g.*, due to division by zero) was an alternative window size chosen. For method 1, the number of standard deviations (for calculation of candidate mask border points) was chosen to be 2.5 and 3.0. For method 2, the scaling factor was chosen to be 0.975 and 1.0. Twelve candidate ID masks were created for each pair of experiments (2 methods, times 3 intensity window sizes times 2 scale or SD factors). For each pair of experiments the mask with the lowest percentage of original data points lying outside of the mask region was selected. The results are shown in table 3.2.8.

The particular values that were obtained for these experiments are only shown for illustrative purposes. For any microarray technology adopted, even a simple analysis of this sort on duplicate experiments will result in much greater accuracy. That is, these masks will provide much greater sensitivity and specificity control than a single arbitrary fold-ratio threshold. For even greater accuracy, this kind of noise modeling must be done on a per-gene basis as described in section 3.3.

| | Exp | | | Exp | | | Exp | |
	A	A	B	B	B	C	C	C
Range Size	1000	5000	9000	2000; 1000	5000	9000	2000; 1000	5000	9000
$\sigma = 3$	3.1	2.7	2.2	93.2	80.5	1.7	99.6	100.0	2.0
$\sigma = 2.5$	11.1	3.8	3.3	97.3	97.6	2.7	100.0	100.0	2.4
$\sigma = 1.0$	19.2	6.2	0.7	100.0	99.0	2.4	100.0	100.0	2.4
$\sigma = 0.975$	19.2	6.2	0.7	100.0	99.3	2.4	100.0	100.0	2.9

Table 3.3
Performance of the 12 identity masks. Each pair of identical experiments gave rise to 12 candidate ID masks. Six of these 12 were derived by method 1 (three with 3 SD and three with 2.5 SD). The other six were derived by method 2 (three with scale 1.00 and three with scale 0.975). Shown here are the percentages of original data points lying outside of the mask region for each of the 12 candidate ID masks derived for experiments A, B, C. $\sigma =$ standard deviation; intensity range or window size of 2000 instead of 1000 is used in experiments B and C for the method 1 trials. (Derived from Tsien *et al.* [181].)

3.3   Prototypical Objectives and Questions in Microarray Analyses

We will now discuss common analytic approaches and issues surrounding the basic problems of finding similarities and differences in reported gene expression levels between the following (refer to table 3.3):

S1. Genes in a microarray experiment, *e.g.*, *intra*-array comparisons of the expression levels in experiment E_m of genes G_k versus G_r in table 3.3, or between genes G_k and G_r in experiments E_m through E_y.

S2. Experiments for a gene, *e.g.*, *inter*-array comparisons of the expression level of G_k in experiments E_m versus E_y, or between experiments E_m and E_y across all genes.

Note that for expositional clarity, we are assuming a one-to-one correspondence between an experiment and a microarray assay.

As has been noted previously, the two most prominent characteristics of microarrays are that they enable the parallel assay of numerous RNA species levels, and the irreproducibility of, and *noise* in, their reported expression levels. In view of these characteristics, the principal objective of microarray data analysis is to extract or discover *knowledge* about a biological system from the wealth of gene expression *information* that is obtained in the wake of each set of chip experiments. Specifically, *information* here primarily consists of noise-ridden gene expression data which

		$\cdots$	Experiment E_m	$\cdots$	Experiment E_y	$\cdots$
$\vdots$						
Gene G_k			x_m^k		x_y^k	
$\vdots$						
Gene G_r			x_m^r		x_y^r	
$\vdots$						

Table 3.4
A prototypal microarrays experiment data set.

may be pangenomic and which may be supplemented by *a priori* known biological facts. *Knowledge*, on the other hand, is certain aspects of the system under investigation that can be elucidated by gene expression data and which includes correlations between genes or conditions, genes whose expression levels distinguish dissimilar biological states, or, at a finer level, regulatory mechanisms and directions of causality between genes, or between conditions and genes. At this point in time, the question of which biological aspect can be unraveled by expression data has neither been entirely explored nor clearly defined.

In the posthybridization analysis, one will inevitably perform the basic *intra-* and *inter*-array comparisons **S1** and **S2**. We encounter situation **S2**, for instance, in studies that seek to find gene markers for biological or physiological states A versus B, where experiments E_1 through E_{m_A} are m_A replicate expression measurements in state A, and E_{m_A+1} through $E_{m_A+m_B}$ are m_B replicate measurements in state B. A time course study where experiments E_1 through E_m are the expression levels sampled at m time points $t_1, \cdots, t_m$, is a typical scenario where **S1** is used. Furthermore, both **S1** and **S2** could apply in a study where experiments E_1 through E_{2m} are m different temperatures sampled twice per temperature condition. As mentioned in section 3.2, with the irreproducible nature of microarray reported measurements in mind, any attempt to reach the above objective in a consistent and rigorous manner will first require well-defined and mathematically workable notions of reproducibility and similarity of gene expression levels as reported by microarrays.

These definitions should capture or abstract, but not contradict, the biological experience, an immediate corollary of which would be an empirical approach to determine the *significance* of a reported gene expression level change. Since the full details of the thermodynamical or physical states of a chip experimental system cannot yet be practically obtained, nor can the sum total of all possible noise

sources be characterized, any definition of reproducibility would typically have to be predicated upon preanalysis assumptions about the behavior of noise in that system.

3.3.1   Two examples: *Inter*-array and *intra*-array.

For heuristic purposes, we will describe inter-microarray and intra-microarray analysis situations which capture the most essential related features of **S1** and **S2** from the introduction of this chapter. While these examples only involve Affymetrix oligonucleotide microarrays, the analytic approach and problems are more general in scope and the methodologies can generally be extended to many other gene expression assaying technologies following easy domain-specific modifications.

Study S1: Yeast cell cycle. A mycologist wishes to find groups of yeast genes whose expression profiles are similar over a 24-hour period. For this, she or he obtains gene expression measurements using Affymetrix S98 yeast genome microarrays for a synchronized sample of yeast cells over a 24-hour period by sampling the total RNA from this population at 30-minute intervals. A total of 48 separate time points were sampled twice (for duplicate measurements) at each time point: $T_1, \cdots, T_{48}$ (and replicates $T'_1, \cdots, T'_{48}$).

Study S2: A drug intervention study. A pharmacologist wishes to characterize the effect(s) of drug X 3 hours after it has been introduced into normal adult wild-type mice by the expression levels of liver cell genes that can be probed using Affymetrix Mu11K microarrays. The gene expression profiles of normal adult mice liver cells that are not treated with drug X are used as the control state. Call the preintervention or control state A, and the postintervention state B. For replicate measurements, liver samples were obtained without drug X application from M_A adult mice and another M_B adult mice liver samples were obtained after drug X was applied.

Detailed specifications of the Affymetrix technology are available in the Affymetrix GeneChip analysis manual [10]. A technical outline of this technology, namely on the interpretation of `.cel` and `.chp` files, is found in section 3.1.2. Affymetrix Mu11K murine chips, *e.g.*, have total probe sets for $N = 13{,}179$ unique transcripts, including "housekeeping genes," whereas the yeast genome S98 array has $N \approx 6400$.

The study properly begins at the experimental design stage: Mice liver explants in study **S2** are harvested and prepared under identical conditions—these most obviously being similar time of day, environmental conditions, and precautionary procedures taken to minimize the effects of non-X-related artifacts and noise in the

gene expression, and *mutatis mutandis* for the yeast cells in study **S1**. Controlling for extraneous factors, which are potentially microarray-detectable in *inter*-array study **S2**, is clearly more difficult and some other factors that should be considered include the age, physical condition, and sex of the mice candidates. Despite the most stringent replicate protocol, it is clear that there always exists some biological or biochemical variation between any two mice, any two cells taken from a common host, or even between any two synchronized yeast cells. In practice, one can only try one's best to exclude explants from mice whose observed or known physiological state is believed to be associated with a different response pattern to X and which might skew the experimental data significantly. With the obvious modifications, whole yeast cells are harvested and duplicates are collected similarly at each of the 48 time points. Following a well-designed analysis scheme, one would hope to be able to classify such negligible variations as noise. To this end, it is important that replicate experiments be run for any chip study. Replicate measurements allow one to quantify the extent to which variations were detectable by microarrays, and to decide whether this noise will confound the reported measurements and attendant analyses. Study **S2** has M_A and M_B replicate experiments for each state. Chapter 2 presents a comprehensive treatment of an experimental design in the presence of noise.

For the remainder of this chapter, we will primarily be concentrating on *inter*-array study **S2**, and developing the mathematical tools, *e.g.*, normalization techniques and measures of (dis)similarity which will be used in chapter 4 to address analyses which are essentially *intra*-array such as study **S1**. We note, however, that most of the questions and methodologies in *inter*-array analyses are equally relevant to the *intra*-array instance, with the obvious contextual modifications. In analyses involving Affymetrix oligochip data, the .chp file is the usual starting point for a typical microarray analysis, (see section 3.1.2). From this file, we will be considering the Probe Set Name, Avg Diff and Abs Call columns. Consolidating the columns of the M_A, M_B .chp files from study **S2** into a table, we have table 3.3.1:

Let A_j and B_j denote the Avg Diff value corresponding to a microarray experiment before and after drug X intervention, respectively, with duplicates $j = 1, 2$. Note that the replicate hybridizations $A1$ and $A2$ might have been scanned into image files using the different user-defined settings or under different ambient conditions beyond the control of the experimenter. Therefore, the most natural question that one asks in any inter-chip study is whether one can validly compare expression levels, represented by the Avg Diff, for any one gene across .chp files that correspond to separate microarray experiments. For instance, for Gene 9785 in

Probe Set Name	A_1		A_2		B_1		B_2	
Gene 1	64.3	(P)	248.2	(P)	128.6	(P)	124.1	(P)
$\vdots$								
Gene 9784	1211.1	(P)	1250.0	(P)	-4.6	(A)	18.3	(A)
Gene 9785	250.5	(P)	193.3	(P)	110.7	(P)	141.2	(P)
Gene 9786	-154.3	(A)	-63.1	(A)	1.1	(A)	-22.0	(A)
$\vdots$								
Gene 13,179	-40.9	(P)	-107.7	(A)	459.7	(A)	198.8	(A)

Table 3.5
Avg Diff (Abs Call) data for *inter*-array study *S2* where $M_A = M_B = 2$

table 3.3.1 we may ask:

Q1. Do order relations (*e.g.*, $<, \geq, =$) of reported Avg Diff values for Gene 9785 in different experimental conditions imply comparisons between the true expression level of the measured gene under these conditions? In particular, while one can convincingly argue that the expression level of Gene 9785 is less intense than the level for Gene 9784 because the *intra*-chip Avg Diff of Gene 9785 (250.5) is smaller than that of Gene 9784 (1211.1) in experiment B_1—assuming that all probes are equally effective—it is not immediately clear whether it is reasonable to claim that the true expression level of Gene 9785 in condition B_2 is greater than in condition B_1 even though the *inter*-chip numerical Avg Diff of Gene 9785 in B_2 (141.2) is greater that the Avg Diff in B_1 (110.7).

Q2. How would one characterize or distinguish the expression (*e.g.*, fold) change from pre- to postintervention states for Gene 9785 that is *most likely* due to a drug X intervention from an expression change that is a caused by measurement variations or, more broadly, noise?

Note that in the ideal system, where every chip assays every transcript equally well and the experiments have been designed so that only expression changes which are a direct consequence of drug X intervention are recorded by the chips with all other attendant thermodynamical or physical variations being negligible and not microarray-detectable, these questions possess trivial answers. Namely, we expect that being replicates, the Avg Diffs $B_1 = B_2$ and $A_1 = A_2$ for each gene, and assuming that A_j, B_j are positive-valued, B_j/A_j quantifies fold change for each gene after intervention. Such utopian scenarios do not appear to be achievable at any time in the near future. In the following sections, we describe several ways to address the two preceding questions under different assumptions and conditions that are commonly used in the bioinformatics literature.

3.4   Preprocessing: Filters and Normalization

We will be referring to the data from study **S2** as tabulated in table 3.3.1. Since the number of replicate experiments in a typical microarray study does not normally exceed three, it is not easy to derive meaningful statistics for any one gene "latitudinally" across two or three repeat measurements, *e.g.*, the mean average reported intensity for Gene 9785 in state B is $(250.5+193.3)/2 = 221.9$, Descriptive statistics such as means or variances are generally informative of the underlying stochastic process in the system being investigated only when one has a large relevant data set to work on. Even though each microarray assay yields one noisy reading for each of the $\sim 10^4$ individual genes, the reported measurement for gene k will not inform us about the robustness of the measurement for gene j $(j \neq k)$, in the absence of at least one other whole replicate assay.

On the other hand, even prior to any transformation(s) that will render the distinct replicate data sets comparable to each other, we can obtain informative statistics for each chip experiment data set "longitudinally" or *intra*-array across all genes 1 through N such as the following:

P1. The mean, standard deviation, minimum, and maximum of Avg Diff in each microarray experiment. For example, for chip A_2 of study **S2** we could find that the mean is 232.4, standard deviation 888.1, minimum -2000.7, and maximum 19792.8.

P2. The mean and standard deviation of Avg Diff differences for all duplicate pairs. For example, the average and standard deviation of $(B_2 - B_1)$ for all 13,179 genes in study **S2** may be -35.4 and 200.3, respectively. This tells us that, on average, a gene registers a lower expression level in B_1 in contrast to experiment B_2. The standard deviation may be informally regarded as an inverse indicator of reproducibility of the replicate readings. This is not strictly an *intra*-array calculation.

P3. The distribution range of Avg Diff values on each chip.

P4. The distribution of genes which have J $(J \leq 4)$ many Present (P) Abs Call. For instance in study **S2**, 2547 genes have a P call in all experiments, 4112 genes have Absent (A) calls throughout all experiments, and 3792 genes have exactly three A calls in all four experiments.

Gross overall data statistics like the above can quickly be used to detect systematic *inter*-chip differences. For instance, if the genomicist in study *S2* finds that for *P2*, the average replicate difference $(B_1 - B_2)$ was 11714.4, then he or she might have good reason to suspect that chip B_1 was scanned at a much higher ambient or background setting than was B_2. Such statistics have been used as preliminary and very crude filters to reduce the number of potentially interesting candidate genes in some microarray studies. For instance, the genomicist in study *S2* could decide that she would only consider (following *P4*) the 2547 genes which have P Abs Calls throughout all 4 experiments for more refined data analysis. Alternatively, the genomicist could accept only genes whose Avg Diffs in some or all of the four experiments are above a predetermined cut-off value. These kinds of filters have their weaknesses. In table 3.3.1, we clearly see that Gene 9784 would be missing from the list of potentially interesting candidates for study *S2* following the all-P-Abs-Call filter, and yet from a cursory inspection of the raw data, and in the absence of further analysis, it seems that drug X triggers a significant reported expression decrease in Gene 9784 from its control levels.

In general, any gross preliminary filtering of the sort above leads to a loss of information. Loss of information is inevitable in typical microarray analyses and is not always a negative feature. After all, microarray data analyses are essentially reductionistic by nature due to the problems of irreproducibility as well as cost limitations in confirming candidate genes resulting from these analyses. The relevance of the loss is context-specific. It depends upon the sort of information that is lost and the scientific question being investigated. Gross statistics such as *P1* and *P3* have also been used to "scale" or *normalize* one chip data set against another so that transformed numerical indicators of expression intensity (Avg Diffs) are *comparable* in some probabilistic sense. This is the gist of question *Q1*.

3.4.1 Normalization

We shall use the term *normalization* to, informally, refer to the transformations that are applied onto data sets from chip experiments to render them *comparable* to one another in a probabilistic or statistical sense, or to take into quantitative account assumptions about how these data were generated. Normalization of a collection of different data sets is typically carried out with respect to a reference data set. The reference set could either be data from one particular experiment, or a postulated distribution of expression intensities. The following are three normalization methods in common practice:

N1. Linear regression

N2. Probabilistic distributions: first, second or k-th moments

N3. Housekeeping genes, and spiked targets

The choice of transformations will depend upon assumptions that are made about the properties of individual data sets such as intensity distribution and the behavior of noise. This will subsequently influence the characterization of expression or fold change significance at later stages in the analysis.

Linear regression   Our discussion will first focus upon the assumptions correctable by linear transformations, ***N1***. Referring to study ***S2***, let x_i and y_i denote the reported expression intensity (Avg Diff) of Gene i in the duplicate experiments B_1 and B_2 respectively with $i = 1, \cdots, N$, where $N = 13{,}179$ is the total number of unique probes on the Mu11K chip set. Let B_1 be the reference data set. For notational clarity, let $\vec{x}$ denote the vector of gene-ordered intensities in a chip experiment, $(x_1, \cdots, x_N)$ in $\mathbb{R}^N$, and let $\bar{x} \doteq \frac{1}{n} \sum_{i=1}^{N} x_i$ be the mean of the gene expression intensities x_i in the chip set B_1.

Now consider for the moment a hypothetical situation where chip B_2 has a systematic error in relation to its (reference) duplicate B_1 such that the Avg Diff for every Gene i, x_i in B_1 is remeasured or reproduced in the B_2 experiment as $y_i = \alpha_1 x_i + \alpha_0$ for some real constants α_0 and α_1. A global linear shift of this kind could arise, for instance, if chip B_2 were scanned at a different uniform ambient brightness from chip B_1. Physically, α_1 is the magnification or *dilation* factor and α_0 is the *translation* factor. A scatter plot of x_i versus y_i would then look like a discretized line of slope α_1 with its vertical or y-intercept at α_0, provided that not all the x_i are equal to a constant. Clearly, if chips B_1 and B_2 were ideal duplicates, then the points (x_i, y_i) would be aligned on a line of slope one through the origin.

This graphical representation of our supposedly replicated raw data suggests an intuitive way to resolve the hypothetical problem. Knowing the values α_1 and α_0, the transformation $y_i \mapsto \hat{y}_i \doteq (y_i - \alpha_0)/\alpha_1$ will correct this systematic difference in B_2 with respect to B_1 so that a plot of x_i versus $\hat{y}_i$ is a line of slope one through the origin $(0, 0)$.

In the general situation, one usually has no *a priori* knowledge of the slope α_1 and y-intercept α_0 of the x_i versus y_i scatterplot. Instead, the main task in this case is to determine the values α_0 and α_1. Normally, one supposes that the data set B_2 is most likely to be systematically different from B_1 by a linear transformation just described, and assume that with B_1 and B_2 being duplicates that their scatterplot should ideally approximate, in the sense of *least squares*, a line of slope one through the $(0, 0)$. Using standard linear regression techniques, one computes α_0 and α_1 for the equation of the line which passes through the data (x_i, y_i) and which minimizes the sum of squared vertical or y-errors between this line and the observed data points (x_i, y_i). In other words this line $f(x) = \alpha_1 x - \alpha_0$, say, should minimize $\sum_{i=1}^{N} (y_i - f(x_i))^2$. Solving for α_1 and α_0 using simple calculus, linear algebra gives us the dilation and translation factors,

$$\alpha_1 = D^{-1} \begin{vmatrix} \sum x_i y_i & \sum x_i \\ \sum y_i & n \end{vmatrix}, \quad \alpha_0 = D^{-1} \begin{vmatrix} \sum x_i^2 & \sum x_i y_i \\ \sum x_i & \sum y_i \end{vmatrix},$$

$$D = \begin{vmatrix} \sum x_i^2 & \sum x_i \\ \sum y_i & n \end{vmatrix},$$

$$(3.4.1)$$

where $|M|$ indicates the determinant of the square matrix M. The normalization transformation then is $y_i \mapsto \hat{y}_i = (y_i - \alpha_0)/\alpha_1$. As an additional bonus, note that the averages of the normalized set B_2 and the reference set B_1 are equalized, i.e., $\frac{1}{N}\sum_{i=1}^{N} \hat{y}_i = \frac{1}{N}\sum_{i=1}^{N} x_i$. Verify this. Question: Are their respective standard deviations the same?

Variations of **N1** that have appeared in the literature include using a subset of genes for the regression calculation as, for instance, using only genes which have present Abs Calls in both the B_1 and B_2 sets, or excluding data points (x_i, y_i) when either x_i or y_i lie below a threshold. Furthermore, if one supposes that the microarrays assay a large enough number of different genes and that the subset of these genes whose expression levels change significantly between conditions A to B is small compared to the total number of genes being measured, then one may also also normalize pairs of experiments A_j, B_j which are not true biological replicates. A useful exercise: Let us say that we have normalized data A_1, A_2 via **N1** with respect to B_1 to produce $\hat{A}_1$, $\hat{A}_2$ separately. Is the linear regression of data $\hat{A}_1$ versus $\hat{A}_2$ a line of slope one through $(0,0)$? What happens when $\hat{A}_1$ and $(\hat{A}_1 - \hat{A}_2)$, regarded as ordered vectors in $\mathbb{R}^N$ are orthogonal to one another?

Probabilistic distributions The second category of normalization schemes **N2** operates under the premise that the distribution of expression levels or Avg Diffs in each chip experiment, or more stringently for duplicate sets of experiments, should be identical. Consider again the data sets $B_1 = \{x_i\}_{i=1}^{N}$ from table 3.3.1. Several studies in the bioinformatics literature have applied the standard central limit theorem-type transformation, $x_i \mapsto \hat{x}_i \doteq (x_i - \bar{x})/\sigma_x$, on each experimental data set. Here, σ_x denotes the standard deviation for the B_1 data set, so that the transformed data B_1 will now have mean 0 and variance 1. Exercise: Consider another data set $B_1 = \{y_i\}_{i=1}^{N}$ that has been normalized similarly $y_i \mapsto \hat{y}_i \doteq$

$(y_i - \bar{y})/\sigma_y$. Show that the linear regression of $\hat{x}_i$ versus $\hat{y}_i$ is a line of slope $\sum_{i=1}^{N} \hat{x}_i \hat{y}_i$ $\left(= N^2 \mathrm{Cov}(\hat{x}, \hat{y})\right)$ through the $(0,0)$. Here $\mathrm{Cov}(\cdot, \cdot)$ is the symmetrical covariance function.

Note that this approach of setting the first two moments of each data set to constants does not guarantee that all the transformed sets of intensities have the same distribution, unless the original data sets were gaussian-distributed to begin with, in which case they would be completely characterized by their first (mean) and second (variance) moments. As an aside, it may be useful to check whether an arbitrary data set is indeed gaussian-distributed. We do this qualitatively with a quartile-quartile plot between the test data set and another data set of the same cardinality that is known beforehand to be gaussian-distributed. A gaussian-distributed data set can be generated using the pseudo-random number utility on most computing machines [144]. These two data sets are first individually ordered by magnitude and then plotted pairwise; for example, if these ranked data sets of cardinality n looked like $\{s_i\}_{i=1}^{n}$, $\{t_i\}_{i=1}^{n}$, then we plot (s_i, t_i). If B_1 is gaussian-distributed, then its quartile-quartile plot against a gaussian-distributed reference set will be linear.

On equating the statistical moments of any two data sets, the reader should also note that even if two distributions share the same kth moments for all integral k, it does not imply that the two distributions are the same (or in the probabilistic parlance, almost surely equal). As a counterexample from Casella and Berger [42], the distributions with probability density functions $f_1(x) = \frac{1}{\sqrt{2\pi x}} \exp[-(\log x)^2/2]$ and $f_2(x) = f_1(x)[1 + \sin(2\pi \log x)]$ have k moments that agree; we clearly see from their graphs that these are different distributions. Furthermore, Press et $al.$ [144] have observed that it is not uncommon to encounter real-life data sets that have finite means but arbitrarily large second moments. The reader ought to be aware of the existence of such pathological cases. These cases are not central to our discussion and the interested reader is referred to [144] for details. Less stringent variations of this normalization scheme are also in common use; $e.g.$, methods which equalize only the first moments (means) of intensity data sets as, for instance, $y_i \mapsto \frac{\bar{x}}{\bar{y}} y_i$, or $y_i \mapsto y_i - (\bar{y} - \bar{x})$, where the mean of the data set $\{y_i\}_{i=1}^{n}$ is transformed to coincide with the mean of a reference set $\{x_i\}_{i=1}^{n}$, by a dilation or a translation.

Housekeeping genes and spiked targets The $N3$ category of normalization devices works on the assumption that certain probes or targets have a known constant hybridization behavior throughout all experimental conditions. For example, exact amounts of particular mRNAs are incorporated or spiked into the target that have a previously known and deterministic effect on specific probes on the microarray under any condition. These special genes could function as housekeep-ing devices in any chip experiment. In this context, the normalization technique typically amounts to transformations on whole data sets, which will narrow or elim-inate the statistical divergence in expression intensity values or Avg Diffs of these housekeeping genes across experiments.

General assumptions and principles of normalization The normalization methods that are outlined above postulate a linear and systematic nature in mi-croarray measurement errors, and assume that every target-probe complex behaves in a similar manner, *i.e.*, hybridization rates are equal and independent of the transcript sequence. More general normalization techniques which are nonlinear or nonparametric have been described [156, 182]. Alternative normalization techniques for microarray data sets in the literature have included the use of eigenvectors [5], a scatterplot smoother [61], normalizing by both sample and gene [81], and mapping the expression data into a real interval between 0 and 1 [193].

Again, we emphasize that the choice of normalization procedure depends entirely upon the postulates, particularly in the noise-related assumptions that one makes for a set of microarray experiments. Recall that we had originally visited the topic of normalization with the aim of resolving *Q1*, *i.e.*, whether order relations (*i.e.*, $\leq, >, =$) between *inter*-chip Avg Diffs for a gene *as is* imply a comparison of the true *inter*-chip expression level of a gene across different chip experimental conditions. While normalization techniques like *N1* attempt to correct global and *deterministic* measurement errors which have the form of a linear transformation, it does not resolve the *stochastic* component of the error, in other words, error or variations due purely to *chance*. Chance-type errors are often subsumed under the term *noise* and are almost always modeled as a stochastic process. Graphically, stochastic effects are reflected by the scatter pattern of data points about some line of regression, as one sees in figure 3.15. In accounting for noise, the genomicist would often makes assumptions regarding the stochastic model. For instance, she or he could postulate that the repeat measurements are random variables of intensity-dependent gaussian distributions whose means lie along the regression line and possess a uniform standard deviation. Methods like *N2* implicitly assume that each microarray data set is a set of samples drawn from a probabilistic distribution which is strongly characterized by its first k-moments. This and other modeling assumptions of the noise effect can naturally be used to determine the significance of a change in the expression level of a gene which we will discuss in an upcoming section on fold significance. The reader is advised that even after applying any one of the normalization techniques above, it might not make sense to extrapolate order relations of the true expression levels across experiments or conditions from order relations on the numerical microarray data.

In view of this, one might begin to ask whether there is a weaker notion of inter-chip comparison. That is, while we cannot definitely claim that the true expression level for Gene 9785 under condition B_2 is greater than in B_1 because the Avg Diff value for Gene 9785 in B_2 (141.2) is greater than in B_1 (110.7), can we derive meaningful biological comparisons of expression levels from the data pair (110.7, 141.2) at all, or at least within some statistical or probabilistic framework?

Consider the scatterplot of the replicate data sets B_1 and B_2 with points (x_i, y_i) where $x_i \in B_1$ and $y_i \in B_2$. Assume that the replicate measurement for any gene is a random variable which is characterized by gaussian distributions whose means are expression intensity dependent and have a uniform standard deviation along the line of regression. In this situation, it is reasonable to expect the data points to be mainly clustered within a 1-standard deviation envelope of the regression line. The *outliers* or data points which lie outside or away from this 1-SD envelope may be interpreted as genes whose expression intensity report or Avg Diff reproduces poorly across replicate experimental conditions. So, even though an order relation in the Avg Diff values does not imply a same-order relation between the corresponding true expression levels, the coordinates of the data (x_i, y_i) can be an indicator of reproducibility of a microarray assay depending upon the postulated underlying noise distribution in the system. Next, we consider the scatterplot for the data set pair (B_1, A_1), assume as previously that measurement errors are gaussian distributed about a regression line, and similar to the above we define the outliers as data points that lie outside the 1-SD envelope away from the regression line. Then a naïve interpretation of the outliers in the (B_1, A_1) plot which are not outliers in the (B_1, B_2) plot is that these are genes that have a expression change in going between states A and B that is less likely to be due to measurement error or noise. We can also naturally assign a magnitude to this expression change by considering the number of standard deviations the outlier in the (B_1, A_1) plot lies away from the regression line. This subject rightly belongs under the topic of gene expression change or fold significance which is explored in the next section.

3.5   Background on Fold

In this section, we describe several approaches to resolving the second question $Q2$ posed in subsection 3.3.1. That is, how would one characterize an expression level or a fold change between different experimental states, for instance, pre- versus postintervention, in our ongoing example study *S2*, that is most likely or significantly due to changes in the biological state of a target rather than to measurement error or noise.

Most methods that have been presented in the bioinformatics literature assume that, with all probes being equally effective, a gene which has a statistically larger change in expression level is more likely to have that change due to a difference in biological state rather than to noise. It should be noted, however, that unless we have enough replicate measurements to be able to determine the robustness of a gene expression reading, we cannot entirely rule out the possibility that noise may cause large changes in reported expression levels. Conversely, not all small changes in expression level are produced by noise alone and, in fact, gene expression changes which are below a preset threshold of noise might very well trigger significant physiological effects in a system. Typically, biological states are macroscopic in scale, and transitions from one state to another are, arguably, continuous phenomena, whereas noise occurs at microscopic scales and is discontinuous.

For an informative contrast and guide as to how one might proceed in microarray fold analysis, let us briefly review how fold change for a particular mRNA transcript is determined using the traditional Northern blot. Typically in a blotting run, one starts by isolating total RNA on the order of 10 μg per gel lane from the system of interest. Each gel lane corresponds to an experimental or replicate condition. Depending on the gel lane to be loaded, the sample load may be spiked with standard controls such as glyceraldehyde-3-phosphate dehydrogenase (GAPDH) for background normalization or loading control calculations. The gel loading step is usually performed manually in approximately uniform quantities. An electric current is passed through the gel complex and by electrophoresis, RNA fragments will move at different speeds in the complex according to their size. Afterward, separated RNA is transferred from the gel onto nitrocellulose membrane to produce a blot. Labeled probes are constructed with radioactivity-labeled nucleotides and are designed to be specific to each RNA. The control and test RNA samples are incubated with the probes. The blot is then washed to remove probes that have not specifically hybridized, exposed to an X-ray film, its image scanned, and the intensity *fold* change quantified by a molecular or phosphor imager.

In each gel lane, the intensity of the mRNA of interest is first normalized with respect to its loading control probe intensity. Rather than an absolute and precise quantification, it would be more appropriate to describe the task of the phosphor imager as fold estimation. In the scientific literature, it is common to quote integer-valued fold changes such as 2 or 100. It is rare to find reports of 1.01- or 10.307-fold in the literature. For the average biological scientist, these numerical fold values have a qualitative rather than a precise quantitative meaning. At each stage of the blotting protocol, many occasions exist for error and noise to enter into the analysis and subsequently into the final scanned image and quantification, in ways that are not entirely understood or avoidable, that would render a 10^{-2} precision of folds biologically redundant or meaningless unless one possesses the error statistic for the blotting and imaging protocol. Another factor governing fold precision is the machine sensitivity of the imager. Recall the setup for a typical microarray analysis. As observed previously, a chip experiment may be regarded as the *reverse* of performing $\sim 10^4$ different Northern blots in parallel on a common two-dimensional substrate of area $\sim 10^{-4}\mathrm{m}^2$ of probes upon which one hybridizes the labeled target. Each microarray probe set would correspond to its associated mRNA sample in one gel lane without the loading control and, in the case of Affymetrix oligochips, with an additional set of mismatched probe controls for nonspecific hybridization. Each whole chip has its set of control or housekeeping probes corresponding to the loading control in a gel lane of the Northern blot.

In view of their similarities, one may reasonably expect any noise and reproducibility issue affecting a Northern blot to manifest itself as prominently in a microarray experiment. Due to the microscopic dimensions of each probe feature on a microarray, small irregularities during scanning or fabrication could lead to discontinuous and possibly contradictory outcomes that might not be immediately or practically detectable from among 10^4 other distinct probes. It is equally instructive to reexamine the starting points for the representative fold calculations in both the Northern blot and microarray technologies.

For microarrays, fold analysis starts off from the level of a text file (*e.g.*, a .chp file in the case of the Affymetrix technology) which contains information such as a probe identifier and corresponding indicators of the level of sample probe or transcript that was detected. Theoretically, this text file is a numerical representation of the scanned chip image, specifically of its detected levels of different RNAs, and is generated by the chip manufacturer's software program which implements (often) proprietary statistical image-processing algorithms that are typically opaque to the user. There is always some loss of information in going from the true image to its machine image file, and to the numerical representation of the image file. The relevance of this loss is, of course, context-specific. Normally, in chip data analysis, the bioinformatician will perform statistical tests, classification, or clustering algorithms based on these preprocessed numerical representations of RNA levels alone. Thus the generic chip data analysis conclusions are at least twice removed from the source or microarray image of RNA levels.

In contrast, the end product of a Northern blot analysis is an X-ray film or an image of the blot, and an estimated quantification of fold change for the RNA of interest after the image is processed by a phosphor imager whose working principles might appear to be more transparent and seem closer to the physical phenomenon under investigation. With microarrays, it is possible to begin fold calculations from their image file. First, this approach may require the bioinformatician to develop his or her own image-processing program which might mean that he or she would have to acquire a working knowledge of several specialized disciplines, including image analysis and software engineering, which is a daunting task in itself. Second, the bioinformatician will need to obtain specific information from chip manufacturers about probe identities and their respective coordinates on the array for every different make of chip that he or she will use. Creating this dictionary location-probe feature lookup may be time-intensive, as in the case with some Affymetrix oligonucleotide microarrays where each probe set of 16 to 20 for a transcript is split up and dispersed throughout the physical array.

It should be noted that while the use of the fold or ratio of intensities is a natural way to quantify relative changes in expression level within the framework of Northern blots, it might not be an appropriate relative measure nor might it retain the same meaning when ported over *as is* into microarray data analysis. In calculating the fold change for a gene between conditions A and B as assayed by micrarrays A_1 and B_1, one would very frequently and simply take the arithmetic ratio numerical representation of reported gene expression in the A_1 and B_1 chips. There are problems in doing this. For instance, Affymetrix represents the intensity of a transcript by the Avg Diff which may be a nonpositive number in which case the ratio of nonpositive quantities is physically meaningless. Taking the data from study $S2$ for example, how does one quantify the fold of Gene 9784 from duplicate states B_1 (-4.6) to B_2 (18.3)? Furthermore, the ratio is not defined when the denominator is zero. For clarity, we shall mainly concentrate on examples of fold analyses on the `.chp` text files in the Affymetrix technology of study $S2$.

3.5.1   Fold calculation and significance

The discussions in this section refer to study $S2$. A prototypal question concerns whether the expression level, as represented by Avg Diffs, of Gene i is *significantly* different between states A and B. By *significance* one almost always means *statistical significance*—specifically, a difference that is unlikely to be due to measurement errors or noise, modeled as a random process that is characterized by a postulated probability distribution which is in turn derived either from the replicate data or some *null hypothesis*. The delicate point here lies in deciding on a consistent and well-defined criterion for statistical significance, and more fundamentally on the choice of a null hypothesis.

Similarly to the normalization procedure, diverse approaches exist to account for different assumptions and postulates about the initial condition of the biological system, the behavior of the target-probe complex, and the constitution and properties of noise. As detailed in [144] for a more general applicational framework, these approaches typically start off by computing a set of statistical parameters or statistics from the empirical data such as means, variances, and kth moments which define a test distribution. This is followed by a choice of a suitable null distribution and some calculations to decide where the parameters fall within this null distribution, *i.e.*, comparing between the test and null distributions. If the test statistic falls in a probabilistically unlikely region on the null distribution, then one may conclude that the null hypothesis is false for the data set, which in this context means that the data or test distribution is statistically different from the postulated noise or null distribution. As emphasized in [144], the reader should bear in mind that one can only disprove a null hypothesis, never prove it. That is, the fact that a data statistic falls within a probabilistically likely region of the null distribution does not imply that the fold is equivalent to the postulated noise effect. When the number of samples used for calculating a statistic is reasonably large, a difference of means that is less than the standard deviation may be significant. Whereas when samples are sparse, a difference of means that is much larger that the standard deviation may not be significant. This fact is relevant considering that the generic microarray study typically uses a small number of expression data per gene, corresponding to different experimental or replicate conditions, for calculating the statistic of the gene.

From study **S2** suppose that the data sets $\{A_j\}_{j=1}^{M_A}$ and $\{B_j\}_{j=1}^{M_B}$ have been suitably normalized following any of the methods outlined in the last section. In order to avoid the clutter of notation, we will use the same symbols A_j, B_j to denote the un-normalized and normalized data sets; the assignments should be clear from the context. Let A_j^i denote the Avg Diff for Gene i as assayed by the microarray A_j. Define B_j^i likewise. First, there are several ways to compute the fold statistic, *e.g.*, the mean, of the Gene i ($i = 1, 2, \cdots, N$):

F1. Average of A_j over the average of B_j: $\mu^{\text{Gene } i} = \dfrac{\frac{1}{M}\sum_{j=1}^{M_B} B_j^i}{\frac{1}{M}\sum_{j=1}^{M_A} A_j^i}$

F2. If $M_A = M_B = M$: $\mu^{\text{Gene } i} = \frac{1}{M}\sum_{j=1}^{M} \log\left(\dfrac{B_j^i}{A_j^i}\right)$

F3. More generally in **F2**: $\mu^{\text{Gene } i} = \frac{1}{M} \sum_{j=1}^{M} \log\left(\frac{B_{\sigma(j)}^i}{A_j^i}\right)$ where σ is any permutation on the set $\{1, 2, \cdots, M\}$

Some studies have alternatively chosen to compute the fold of Gene i from states A to B as $\frac{B^i - A^i}{A^i}$ instead of $\frac{B^i}{A^i}$ as above. In order to resolve the problem of non-positive-valued Avg Diffs, it is commonplace in the literature, before taking arithmetic ratios, to threshold the Avg Diff values in every chip data set to an arbitrarily chosen minimum positive number, as *e.g.*, in setting all Avg Diffs of less than 10.0 to 10.0. Alternatively, some studies translate all the intensity data in a chip data set so that the minimal translated Avg Diff in each set is positive. For example, if the minimum element in the A_2 data set is -2000.7, then when 2001.7 is added to every Avg Diff reading in A_2, all the translated Avg Diffs will be positive-valued. Note that such solutions will skew the intensity statistic of the microarray data which is not always a desirable thing. For symmetry reasons, the logarithm of folds rather than just the folds alone are averaged. Consider the data for Gene 1 un-normalized: From A_1 to B_1, the intensity of Gene i changed 2.0 $\left(\frac{128.6}{64.3}\right)$-fold, whereas from A_2 to B_2, it changed 0.5 $\left(\frac{124.1}{248.2}\right)$-fold so that on average the intensity of Gene i should intuitively be unchanged, *i.e.*, have a fold change of 1.0. However, it is obvious that the arithmetic average of 2.0 and 0.5 does not equal 1.0. A logarithmic transformation of the individual folds solves this problem, as does the use of the geometric rather than the arithmetic mean of the folds.

As an exercise, the reader should verify that the order of taking logarithmic ratios in **F2** and **F3** does not change the resulting fold, $\mu^{\text{Gene } i}$. After computing the fold of all genes in the data set in any one of the preceding ways, one typically wants to know whether a fold average or a fold distribution is different from, or *statistically significant* in relation to, a postulated *null distribution* which represents the effects of measurement errors or noise. As we have already noted, it is essentially the choice of the *null hypothesis* which distinguishes the different methods for determining fold significance. Some studies associate the null hypothesis with an interval of nonsignificance informally called a noise envelope. The distribution for the null hypothesis has also been calculated from the microarray data set permuted, especially when the number of replicate data points is small. In general, replication improves the estimation of the null distribution. Below, we outline several common null distributions:

G1. A null distribution for each Gene i whose mean is obtained by averaging latitudinally across duplicates, $\mu^{\text{Noise } i} = \frac{1}{2}\left(\left|\log\left(\frac{A_1^i}{A_2^i}\right)\right| + \left|\log\left(\frac{B_1^i}{B_2^i}\right)\right|\right)$

G2. A null distribution for all genes whose mean is obtained by averaging longitudinally across all N genes, $\mu_A = \frac{1}{N} \sum_{i=1}^{N} \left| \log \left(\frac{A_1^i}{A_2^i} \right) \right|$ or $\mu_B = \frac{1}{N} \sum_{i=1}^{N} \left| \log \left(\frac{B_2^i}{B_1^i} \right) \right|$

G3. For each gene row, permutation of the intensity data, for instance, exchange data between conditions $A_1 \leftrightarrow B_2$, *etc.*, and then recalculate **G1.**

A coarse, qualitative method may combine **F1** and **G2** to decide that a Gene i with $\mu^{\text{Gene } i} > \max(\mu_A, \mu_B)$ or $\mu^{\text{Gene } i} < -\max(\mu_A, \mu_B))$ is significantly different, foldwise, from the average fold statistic resulting from noise as calculated from duplicate conditions. Another approach could be to rank the N genes by their $\mu^{\text{Gene } i}$ value from **F2** and to decide that the top and bottom 5% of these ranked genes are significantly changed.

There exist equally diverse non fold-centered methods for determining the significance of a change in expression. The intuitive idea behind all these approaches is to call a change statistically significant if the expression change *inter*-states A and B is maximal, and the expression change *intra*-state within the replicate conditions $\{A_j\}_{j=1}^{M_A}$ and $\{B_j\}_{j=1}^{M_B}$, respectively, is minimal. Again, the choice of a null distribution representing the postulated noise distinguishes these methods. Being non fold-based, traditional parametric and nonparametric statistical tools for analyzing means or variances between data groups such as the χ^2, Student t, Mann-Whitney, and F tests and ANOVA have been applied in the literature. These methods are reviewed comprehensively in [144] and the reader should be aware of the implicit assumptions and underlying null hypotheses in these standard tests prior to application. A drawback to these traditional approaches is that their conclusions are "asymptotic", *i.e.*, they are statistically valid only when one has a large number of replicate data.

At the end of the normalization section 3.4, we had briefly described a way to graphically visualize the determination of significant difference in expression data between A_1 and B_1. To reiterate, on the scatterplot of the un-normalized data sets A_1 versus B_1, we compute and draw the linear regression line. If chip assay reproducibility is reasonably robust and if we assume that the majority of genes do not undergo a dramatic change in expression level in going from state A_1 to B_1, then the data points should cluster close to this regression line which should have slope one and should pass through $(0, 0)$. The envelope of a standard error or deviation spread away from the regression line is our object of interest. This 1-SD envelope may be defined by a pair of lines that lie along a horizontal distance $\left(\frac{1}{N} \sum_{i=1}^{N} \left(B_1^i - f \left(A_1^i \right) \right)^2 \right)^{\frac{1}{2}}$ to the left and right of the regression, $f(x) = \alpha_1 x + \alpha_0$. The intuitive reasoning here is that a data point which is outside this envelope is a statistically *insignificant* event relative to a postulated stochastic distribution, usually gaussian, of the data around the regression line due purely to noise (chance). Such an outlying data point represents, in the reverse context, a gene which has undergone a statistically *significant* expression change from A_1 to B_1.

3.5.2   Fold change may not mean the same thing in different expression measurement technologies

In one common experimental design, DNA microarrays are built by printing or spotting an array of cDNA on a glass slide, which provides gene-specific probes for hybridization to targets. Two different samples of mRNA are labeled with different fluorescent dyes and simultaneously hybridized onto each probe. The data for each probe (gene) consists of two fluorescence intensity measurements, typically red and green, representing the expression levels of the same gene in the red-labeled (with the dye Cy5) and the green-labeled (with Cy3) mRNA samples. Early analyses of the data obtained through these microarrays will identify differentially expressed genes by taking the ratio of red intensity to green intensity, *i.e.*, the *intensity ratio*, and choosing an arbitrary cut-off above or below which the genes will be interpreted as being differentially expressed in the two samples. The crucial point is that one typically does not use the absolute measure of intensity, but uses, instead, the ratio or degree of difference between samples.

In one of the earliest uses of this design, Schena *et al.* [157] called a gene differentially expressed if the expression level in the two mRNA samples differed by more than a factor of 5. DeRisi *et al.* [56] called genes differentially expressed if the logarithm of their ratio differed by more than a factor of 3, after standardizing the intensity measurements using a set of "housekeeping" genes. More recently, sophisticated statistical techniques have been developed to provide more reliable calculations of the intensity ratios by taking into account the distribution of the intensities across the whole microarray. In the first statistical analysis of these data, Chen *et al.* [45] proposed a method to identify statistically significant changes between the two mRNA samples, under several assumptions on the distribution of the intensities, including normality, null intercept, and constant variation coefficient. Newton *et al.* [135] offered a bayesian approach to the problem, using a hierarchical model for intensities and identified differentially expressed genes on the basis of the posterior odds of change. It is worthwhile to mention that Arfin *et al.* [11], in a study involving cDNA arrays expression profiling of *E. coli*, found little correlation between fold difference in the reported expression levels of a gene and its accuracy.

Using synthetic oligonucleotide microarrays, on the other hand, one identifies differentially expressed genes by comparing expression measurements made on two separate microarrays. On these microarrays, there is no simple measure of intensity: instead, 16 to 20 probe pairs are used to interrogate each gene. Each probe pair has a PM and MM intensity signal. In an ideal setting, if a gene is truly present, each of the PM probes would have a high intensity and each of the MM probes would have a near-zero intensity. The ratio PM/MM closely resembles the intensity ratio used with spotted microarrays; however, here we are using this ratio to determine an absolute expression level for this gene. Since each gene is represented by many probe pairs, the average log(PM/MM) is called the *Log Average Ratio*. A probe pair would be considered positive if $PM - MM$ is greater than or equal to a threshold, and if PM/MM is greater than or equal to another threshold (similarly, $MM - PM$ and PM/MM are calculated in considering a probe pair negative). The average of the $PM - MM$ differences for all probe pairs in a probe set is called the *Average Difference*, and this value is used as the absolute amount of expression for a gene.

In order to determine whether a gene expression is significantly different, current experimental protocols require the (possibly repeated) pairwise comparison of two microarrays, each hybridized to each of the different samples being compared. The intensity ratio used in the spotted arrays is thus replaced by the indirectly calculated ratio between the the average difference of each gene in the two different arrays. Typically, a cut-off threshold is arbitrarily selected between 2 and 3 (see, *e.g.*, [97, 198]). That is, the expression level of a gene will be considered differentially expressed in an experiment if the (indirectly calculated) average difference is two or three times higher in one microarray than in the other. However, this threshold has been set as low as 1.7 [117].

The major manufacturer of these microarrays [10] provides software implementing a different measure of differential expression. Let x represent the average difference of a gene on one microarray and y its average difference on the second microarray. For each gene, the level of change across the two microarrays is computed by a weighted ratio:

$$W = \frac{y - x}{\max(\min(x, y), \ Q_c \cdot Q_m)}$$

If $y \geq x$, then $+1$ is added to W, otherwise $W - 1$. Q_c is the maximum noise level of the two microarrays, or the difference in pixel intensities seen in the lowest intensity probes, and Q_m is a constant depending on the resolution of the microarray—2.1 for 50 μm feature microarrays and 2.8 for 24 μm feature microarrays. However, the decision of whether a gene is *significantly* differentially expressed in the two microarrays is based on a complicated interpretation of the change of each of its 14 to 20 probe pairs. Let $(PM - MM)_x$ be the difference between PM and MM for a probe pair in the first array and $(PM - MM)_y$ be the same difference for the same probe pair on the second array. A probe pair shows a significant change (increased or decreased) across two microarrays if:

1. the difference $(PM - MM)_y$ - $(PM - MM)_x$ is higher than an arbitrary threshold called the *Change Threshold*, and

2. the ratio, $\frac{(PM-MM)_y - (PM-MM)_x}{\max(Q, \; \min((PM-MM)_x, (PM-MM)_y))}$ exceeds another arbitrary threshold called the *Percent Change Threshold*, where Q represents the noise on the chip.

Given this definition of a significantly increased or decreased probe pair, a gene is considered significantly differentially expressed based on four aggregate measures:

1. The number of increased or decreased probe pairs, whichever is higher.

2. The ratio of the number of increased probe pairs to the number of decreased probe pairs.

3. The difference or subtraction of the log average ratios of the gene in the two microarrays.

4. The difference in the number of positive probe pairs minus the difference in the number of negative probe pairs, divided by the number of probe pairs.

The decision of whether a gene is significantly differentially expressed between two microarrays is decided by a user-modifiable matrix indexed by the above four results. The user can set thresholds to define the boundaries where each metric changes the decision. Default values for these thresholds have been set by the manufacturer through in-house empirical testing.[5] Since systematic variations are expected to occur between any two microarrays—due to biological noise, changed environmental conditions, or other factors affecting equally all genes on an array—various normalization techniques are used to render comparable the expression levels across the two microarrays. Section 3.4.1 discusses the various techniques used.

It should be clear that there is a great deal of difference in the interpretation of intensity measurements, intensity ratios, and gene expression levels across the two systems. Fundamentally, spotted arrays are currently used to measure a *difference* in transcription levels between two samples, whereas oligonucleotide arrays are used to measure *absolute* transcription levels in a single sample. Even if one discounts the

---

[5]As complicated as this methodology sounds in our description here, we are ignoring the fact that in the actual algorithm, not all available probe pairs are used in these calculations. Probe pairs with an intensity difference significantly beyond the mean intensity difference are considered outliers and are ignored in these calculations.

differences in the quantitative and differencing algorithms, it should be obvious to the reader that the "results" of an experiment involving spotted arrays are going to be significantly influenced by the choice of the reference or control sample, already implying that the resulting values are not directly comparable.

3.6 Dissimilarity and Similarity Measures

In time course studies such as study $S1$, it is natural to talk about how two genes G_k and G_j are (dis)similar to one another with respect to their expression levels sampled in time. In the bioinformatics literature, a gene and its corresponding expression levels are often represented as a data point or a vector in $\mathbb{R}^k$, where k is a positive integer (denoted by $\mathbb{N}$). Each vector component or coordinate slot denotes the expression level of that gene under a different experimental condition with k conditions altogether. A measure of similarity or dissimilarity such as a metric is then defined on $\mathbb{R}^k$ which maps pairs of vectors, representing genes and their corresponding expression information, to real numbers. When the number of genes and the expression data set are very large, as in most microarray experiments, one often asks which genes or groups of genes behave *similarly* under these k different or replicate conditions. In an attempt to answer this question, one resorts to classification and clustering methods, and of particular importance prior to the analysis, one has to define what one means by the term *similar*. Similarity is characterized by the choice of a (dis)similarity measure, such as a metric.

As noted by Jain and Dubes [101], a *cluster method* takes as its input a finite data set $\mathcal{X}$ of at least p ($p > 2$) elements together with a list of finitely many attributes of these data, or, alternatively, it may consist of a measure of similarity or dissimilarity between pairs of objects in $\mathcal{X}$. The output is an internal structure of $\mathcal{X}$, which is determined entirely by the input and *not* dependent on any external criterion. Following [103], even though there is no universal agreement in the field with regard to terminology, when the rules for determining the output are externally defined, the algorithm will be called a *classification* rather than a cluster method. For example, suppose that we have for our set $\mathcal{X}$ microarray data which are represented as points in $\mathbb{R}^3$, *i.e.*, equipped with the standard Euclidean distance, a metric. Then, a cluster method would ideally yield groups of genes which are physically close (in Euclidean distance) to each other, and hence behave similarly over three conditions, forming very tangible visual clusters when plotted. On the other hand, questions such as which genes appear in the first octant of $\mathbb{R}^3$ would fall under the classification category. The distinction between the terms *clustering* and *classification* is not purely of academic interest. Clustering algorithms generally have a higher order of complexity, and are therefore more computationally intensive, than classification methods. We begin the discussion of (dis)similarity measures with several formal and standard definitions. Let $\mathcal{X}$ be any nonempty set.

Definition 1 A function $\rho : \mathcal{X} \times \mathcal{X} \to [0, \infty)$ is a metric if for all $x, y \in \mathcal{X}$:

M1. $\rho(x, y) = 0$ if and only if $x = y$ (definite).

M2. $\rho(x, y) = \rho(y, x)$ (symmetrical).

M3. $\rho(x, y) \leq \rho(x, z) + \rho(z, y)$, for all $z \in \mathcal{X}$ (triangle inequality).

Some examples of common metrics are:

- Let $\mathcal{X} = \mathbb{R}^n$, $n \in \mathbb{N}$. The classical Euclidean distance, $\rho(x, y) = \left(\sum_{i=1}^n |x_i - y_i|^2 \right)^{\frac{1}{2}}$ for any $x, y \in \mathcal{X}$, where $x = (x_1, \cdots, x_n)$, $y = (y_1, \cdots, y_n)$.

- Let $\mathcal{X} = \mathbb{R}^n$, $n \in \mathbb{N}$. $\mathcal{L}^p$-type distances with $p \in \mathbb{N} \cup \{0\}$, $\rho(x, y) = \left(\sum_{i=1}^n |x_i - y_i|^p \right)^{\frac{1}{p}}$ for any $x, y \in \mathcal{X}$. When $p = 2$, we have the preceding Euclidean distance.

- Suppose that $\mathcal{X}$ is the set of real-valued continuous functions on the closed real interval $[0, 1]$. Let $f, g \in \mathcal{X}$, and $\rho(f, g) = \max_{x \in [0,1]} |f(x) - g(x)|$. Exercise: Confirm this.

- The distance between two random variables X and Y, $\rho(X, Y) = H(X, Y) - H(X; Y)$ where $H(X, Y)$ is the joint entropy of X, Y, and $H(X; Y) \doteq H(X) - H(X|Y)$ is the mutual information between X and Y, a symmetrical function. An overview of entropy and information-theoretic concepts is presented in subsection 3.6.2. We refer the interested reader to Cover and Thomas [54], for a detailed review of information theory.

The reader should be warned that there is a widespread misuse of the terminology *metric* in the current bioinformatics literature to include functions such as the linear (Pearson's) correlation coefficient and mutual information. It is a simple exercise to show that these functions do not satisfy the axioms of a metric. Next we define a weaker class of functions than the metric, a *dissimilarity coefficient*.

Definition 2 A dissimilarity coefficient is any function $d : \mathcal{X} \times \mathcal{X} \to [0, \infty)$ which satisfies two axioms:

D1. $d(x, x) = 0$ for all $x \in \mathcal{X}$.

D2. $d(x, y) = d(y, x)$ for all $x, y \in \mathcal{X}$ (symmetric).

A dissimilarity coefficient d is *definite* if it satisfies a third condition:

D3. $d(x, y) = 0$ if and only if $x = y$, $x, y \in \mathcal{X}$

Any metric is a definite dissimilarity coefficient, and note that $1 - r$, where r denotes the linear (Pearson's) correlation coefficient is a definite dissimilarity coefficient. Exercise: Confirm these.

3.6.1 Linear correlation

Given data points (x_i, y_i) for $i = 1, \cdots, n$, it is common to ask whether a *linear* relationship exists between the quantities x_i and y_i, and to what degree are the data linear, at least statistically. The graphical equivalent would be to ask whether the scatter plot of x_i versus y_i looks like (discretizations of) a line. The linear or Pearson's correlation coefficient r is the mathematical device that is most often used to do this,

$$r \;=\; \frac{\sum_{i=1}^{n} (x_i - \bar{x})(y_i - \bar{y})}{\sqrt{\sum_{i=1}^{n} (x_i - \bar{x})^2}\sqrt{\sum_{i=1}^{n} (y_i - \bar{y})^2}} \tag{3.6.2}$$

where $\bar{x}, \bar{y}$ denote the means of x_i, y_i respectively. As an aside, observe that,

$$r^2 \;=\; \underbrace{\frac{\sum_{i=1}^{n} (x_i - \bar{x})(y_i - \bar{y})}{\sum_{i=1}^{n} (x_i - \bar{x})^2}}_{\alpha_1^{(x,y)}} \; \underbrace{\frac{\sum_{i=1}^{n} (x_i - \bar{x})(y_i - \bar{y})}{\sum_{i=1}^{n} (y_i - \bar{y})^2}}_{\alpha_1^{(y,x)}} . \tag{3.6.3}$$

In other words, r is the geometric mean of the slope of the linear regression of x versus y and the slope of the linear regression of y versus x, denoted $\alpha_1^{(x,y)}$ and $\alpha_1^{(y,x)}$ respectively. By the Cauchy-Schwarz inequality,

$$\langle x, y \rangle \;=\; \|x\| \, \|y\| \cos \theta \,, \qquad x, y \in \mathbb{R}^n \,, \tag{3.6.4}$$

where $\langle x, y \rangle \doteq \sum_{i=1}^{n} x_i y_i$ is the standard dot product on $\mathbb{R}^n$, $\|x\| \doteq \langle x, x \rangle^{\frac{1}{2}}$ and θ is the angle between vectors x and y, it can be shown that,

$-1 \le r \le 1$,

$r = 1$, if and only if, $x = y$; x, y are positively correlated,

$r = -1$, if and only if, $x = -y$; x, y are negatively correlated.

Suppose that one regards x, y as random variables with samples x_i, y_i indexed by $i = 1, \cdots, n$. If x and y are stochastically independent, then $r = 0$. The converse is not generally true, as, e.g., when we take $t_i = i \cdot 2\pi/n$, $x_i = \cos t_i$ and $y_i = \sin t_i$ for any integer n. Exercise: Verify this. When r is near 0, x and y are said to be uncorrelated. A review of the concept of stochastic independence may be found in Shiryaev [163]. Note that r has a nonrigorous transitive property such that if x and y, and y and z, are strongly correlated with correlation strength $r_{x,y}$ and $r_{y,z}$, x and z have correlation strength r. As noted by Press et al. [144], while r provides a summary of the strength of the association if the correlation is found to be significant, r is not generally useful for deciding whether an observed correlation is statistically significant. This weakness arises from the fact that the computation (3.6.4) ignores the distributions of x and y. One solution to this problem is to use nonparametric or rank correlation, which details are also found in [144]. It can also be shown that r is more sensitive to small perturbations of data points that are farther away from the centroid of the system than to perturbations near the centroid. This may result in a bias. Linear correlation has been used to find pairwise associated genes across different experimental conditions, as in [63].

3.6.2   Entropy and mutual information

Suppose that we have the expression information for a pair of genes G^x, G^y across n experimental conditions. In the most extreme cases, the expression of G^x and G^y may be completely dependent (if this dependence is linear, this implies that $r = \pm 1$), or independent (which implies that $r = 0$) across these n conditions. Question: Can we measure or quantify the dependence of G^x and G^y levels in this case? Equivalently, how much would the expression data for G^x inform us about the corresponding expression of G^y, and vice versa? In this regard, we may appeal to relevant information-theoretic and conceptual devices that were formulated to solve the fundamental problem in communications of reproducing at the receiver endpoint, either exactly or approximately, a message from the transmitter endpoint. The messages transmitted and received will correspond to the expression data of G^x and G^y in the biological scenario. Suppose that X is a random variable with the set of events or outcomes $\mathcal{F}_X = \{x_i\}_{i=1}^n$, and corresponding probabilities $\mathsf{P}(X = x_i) = p_{x_i} \geq 0$. By definition, $\sum_{x \in \mathcal{F}_X} \mathsf{P}(X = x) = 1$. Following Shannon [160], we want to find a measure $H(p_{x_1}, \cdots, p_{x_n})$ of how much "choice" is involved in the selection of events, or of how uncertain we are of the outcome. Intuitively and reasonably, we will require that H have the following three properties:

E1. H is continuous in p_{x_i}.

E2. If all p_{x_i}'s are equal, *i.e.*, $p_{x_i} = \frac{1}{n}$, then H should be a monotonically increasing function of n. With equally probable events, there is more choice or uncertainty in the outcome when there are more possible events x_n.

E3. If a choice is broken down into two successive choices, the original H should be the weighted sum of the individual values of the later H's. For instance, let $n = 3$ and $\{p_{x_1}, p_{x_2}, p_{x_3}\} = \{\frac{1}{2}, \frac{1}{3}, \frac{1}{6}\}$, and we consider three possible decompositions of a choice of the three outcomes: Then $H\left(\frac{1}{2}, \frac{1}{3}, \frac{1}{6}\right) = H\left(\frac{1}{2}, \frac{1}{2}\right) + \frac{1}{2}H\left(\frac{2}{3}, \frac{1}{3}\right) = H\left(\frac{5}{6}, \frac{1}{6}\right) + \frac{5}{6}H\left(\frac{3}{5}, \frac{2}{5}\right)$.

It can be shown that the only explicit form of H satisfying ***E1*** to ***E3***,

$$H(p_{x_1}, \cdots, p_{x_{n'}}) = -K \sum_{i=1}^{n} p_{x_i} \log p_{x_i},$$

where K is positive constant. We let $K = 1$. We shall call $H(X)$ the entropy of random variable X. Consider the case of a Bernoulli random variable X, *i.e.*, $n = 2$, with $p_{x_1} = p$ and $p_{x_2} = 1 - p$ where $p > 0$. Then, when X is one-dimensional,

$$H(X) = -p\log p - (1 - p)\log(1 - p).$$

Using a base-2 logarithm, H plotted as a function of p is concave, non-negative, symmetrical and has a maximum of 1 at $p = 1/2$ (see figure 3.24).

As noted in [160], H has several interesting properties that substantiate it as a reasonable measure of choice or uncertainty.

H1. $H \geq 0$ with equality if and only if all but one $p_{x_i} = 0$ and the non-zero event has probability 1. Therefore, when we are entirely certain about the outcome H the measure of uncertainty vanishes.

H2. For any number of possible outcomes n, H is a maximum ($\log n$) when all the p_{x_i}'s are equal, *i.e.*, $p_{x_i} = 1/n$. Intuitively, this is the most uncertain scenario.

H3. Suppose that Y is a random variable with the set of outcomes $\mathcal{F}_Y = \{y_i\}_{i=1}^{m}$ and corresponding probabilities $\mathsf{P}(Y = y_i) = p_{y_i} \geq 0$ and $\sum_{y \in \mathcal{F}_Y} \mathsf{P}(Y = y) = 1$. Let p_{x_i, y_j} be the probability for the joint occurrence of $X = x_i$ and $Y = y_j$. Define the joint entropy of X and Y to be

$$H(X, Y) \doteq - \sum_{x \in \mathcal{F}_X, \, y \in \mathcal{F}_Y} p_{x,y} \log p_{x,y} = -\sum_{i=1}^{n} \sum_{j=1}^{m} p_{x_i, y_j} \log p_{x_i, y_j}$$

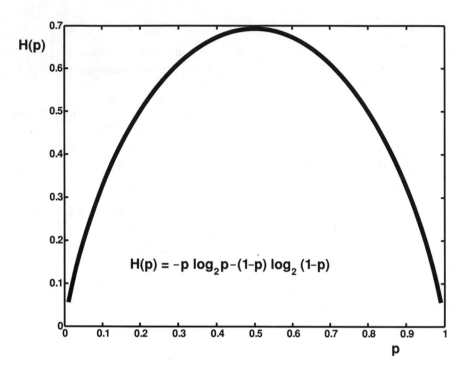

Figure 3.24
Entropy plot

and by basic properties of marginals of joint distributions,

$$H(X) = - \sum_{i,j=1,1}^{n,m} p_{x_i,y_j} \sum_{j=1}^{m} \log p_{x_i,y_j} = - \sum_{i=1}^{n} p_{x_i} \log p_{x_i}$$

$$H(Y) = - \sum_{i,j=1,1}^{n,m} p_{x_i,y_j} \sum_{i=1}^{n} \log p_{x_i,y_j} = - \sum_{j=1}^{m} p_{y_j} \log p_{y_j}$$

Exercise: Show that $H(X,Y) \leq H(X) + H(Y)$ with equality if and only if X and Y are independent, *i.e.*, $p_{x_i,y_j} = p_{x_i} p_{y_j}$. The uncertainty of a joint event is less than or equal to the sum of individual uncertainties. Note that $H(X,X) = H(X)$.

H4. Any change toward equalizing the p_{x_i}'s increases H. Computationally, any averaging operation, $p_{x_i} \mapsto p'_i = \sum_{k=1}^{n} a_{ik} p_{x_k}$, where $\sum_i a_{ik} = \sum_k a_{ik} = 1$ with $a_{ik} \geq 0$ increases H—except in the special case of a permutation where for every i all but one $a_{ik} = 0$, the non-zero constant being 1, in which case H is invariant.

H5. Recall the definition of conditional probability, $\mathsf{P}(X = x | Y = y) = \frac{\mathsf{P}(X=x,Y=y)}{\mathsf{P}(Y=y)}$ if $\mathsf{P}(Y = y) \neq 0$, *i.e.*, $p_{x|y} = \frac{p_{x,y}}{p_y}$. An easy consequence of this definition is Bayes' theorem,

$$p_{y|x} = \frac{p_{x|y} p_y}{p_x} = \frac{p_{x|y} p_y}{\sum_{y' \in \mathcal{F}_Y} p_{x|y'} p_{y'}}$$

The conditional entropy of X given $Y = y$ is defined as

$$H(X|Y = y) \doteq - \sum_{x \in \mathcal{F}_X} p_{x|y} \log p_{x|y}$$

whereas the conditional entropy of X given Y is defined by averaging the $H(X|Y = y)$ over $y \in \mathcal{F}_Y$,

$$
\begin{aligned}
H(X|Y) &= \sum_{y \in \mathcal{F}_Y} p_y \, H(X|Y = y) \\
&= - \sum_{x \in \mathcal{F}_X, \, y \in \mathcal{F}_Y} p_y \, p_{x|y} \log p_{x|y} \\
&= - \sum_{x \in \mathcal{F}_X, \, y \in \mathcal{F}_Y} p_{x,y} \log p_{x|y}
\end{aligned}
$$

This quantity measures how uncertain we are of Y on average when we know X. The chain rule for entropy can then be shown. Exercise:

$$H(X,Y) = H(X|Y) + H(Y) = H(Y|X) + H(X)$$

Now following the exercise in **H3**, we can conclude that,

$$H(X) + H(Y) \geq H(X,Y) = H(X|Y) + H(Y)$$
$$H(X) \geq H(X|Y)$$

which says that *on average* our knowledge of Y does not increase our uncertainty of X. The reader is encouraged to come up with a pair of X, Y such that $H(X|Y = y) > H(X)$ but $H(X|Y) \leq H(X)$. When X and Y are independent, our uncertainty of X is unchanged; otherwise it decreases.

H6. The mutual information between X and Y is defined as

$$M(X,Y) \doteq H(X) - H(X|Y)$$

Exercise: Show that $M(X,Y) \geq 0$ and $M(X,Y) = M(Y,X)$. Mutual information measures the average reduction in uncertainty about X given that we have knowledge of Y, and *vice versa* by the exercise. When X and Y are independent, $M(X,Y) = 0$, *i.e.*, there is no reduction in uncertainty. At the other extreme, when $X = Y$, $M(X,Y) = H(X)$ so that we have reduced all uncertainty—$H(X)$—about the system. Here we also see that M is *not* a metric since $M(X,X) \neq 0$. The distance between random variables X and Y is defined as

$$D(X,Y) \doteq H(X,Y) - M(X,Y)$$

Exercise: Show that D satisfies the three axioms of a metric in an earlier section.

H7. The cross or Kullback-Liebler entropy between the probability distributions p_x and q_x over a common space of possible outcomes $\mathcal{F}_X$ is defined as,

$$D_{\mathrm{KL}}(p,q) \doteq \sum_{x \in \mathcal{F}_X} p_x \log \frac{p_x}{q_x}$$

Note that in general, $D_{\mathrm{KL}}(p,q) \neq D_{\mathrm{KL}}(q,p)$.

In order to apply the preceding tools in the context of microarray data analysis, suppose that we have the gene expression data for genes G^x, G^y across m different experimental conditions. Divide the ranges of the G^x, G^y expression data into uniformly spaced n intervals or "bins." Let p_{x_i} be the probability that the expression for G^x falls into the ith interval, and similarly for G^y. Suppose that three out of the m different experiment intensities for G^x fell within the kth bin, then we let $p_{x_k} = 3/n$. A graphical example of this is shown in figure 3.25.

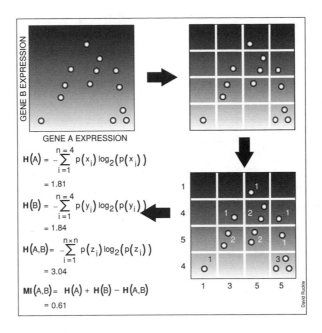

Figure 3.25
Graphical example of mutual information calculation. First, a scatterplot of the expression measurements of the two genes is created, and a grid is imposed. In this example, each expression measurement is quantized into four bins (one can think of these as "low," "low-medium," "high-medium," and "high," though any number and positioning of bins can be considered). The entropy for each gene is then calculated using the row and column sums, and the joint entropy is calculated from the grid.

3.6.3   Dynamics

Dynamics provides a powerful example of how the choice of the appropriate similarity measure can affect the analysis of a genomic system. We use the term *dynamics* to refer to the rate of change of genetic expression over time, calculated as the first-order difference of the genetic expression levels (E_{t2}-E_{t1}, E_{t3}- E_{t2}). This is different from the simple temporal pattern of genetic expression (E_{t1}, E_{t2}, E_{t3}) that we refer to as statics. The primary motivation for studying gene expression dynamics is that existing static techniques may not identify all the important relationships. Some genes may have associated dynamic behaviors but may not have associated static expression behaviors. A hypothetical example is shown in figure 3.26: Gene A codes for an enhancer protein that regulates the expression of Gene B—a high level of Gene A causes an upregulation of expression in Gene B. Since Gene B can be at many possible expression levels before being affected by Gene A, the enhancer-type relationship between the two genes cannot be noticed by simply examining the correlation of static expression patterns. Instead, one needs to examine the dynamics of gene expression—the way in which the expression level of Gene A leads to a change in Gene B—to detect the underlying dynamic relationship. We therefore hypothesized that using dynamics measures as our fundamental similarity measure has the potential to discover relationships between genes that are not detectable using static similarity measures.

Reis *et al.* [150] investigated the *Saccharomyces cerevisiae* mRNA expression data aggregated from several experiments reported by Eisen *et al.* [63] in which the response of the yeast cells to several different stimuli is recorded. The data contain 79 data points in 10 time series measured under different experimental conditions. Of over 6000 genes in the yeast genome, Eisen *et al.* included only 2467 genes that had functional annotations. We analyzed the same subset of genes as described. Dynamics similarity measures were calculated as slopes of the change of expression over time. Slopes were calculated between each adjacent pair of expression data points, E_{t_n} and $E_{t_{n+1}}$: $\text{Slope}_{(n,n+1)} = \frac{\text{Expression\_Level}_{n+1} - \text{Expression\_Level}_n}{\text{Time}_{n+1} - \text{Time}_n}$ As slopes are only calculated between data points within the same time series, the 79 data points in 10 time series are reduced to only 69 slope measurements. The units of the slope measurements are in normalized expression level units per minute. The authors found those pairs of genes that exceed a correlation coefficient of .78 based on a permutation analysis (see section 4.12.1). We compared the pairs of genes found to be correlated at that threshold or higher using both the standard static expression levels and the slope similarity measures. We found that 133 genes appeared in both the static and dynamic analyses, leaving 215 genes that were exclusive to the

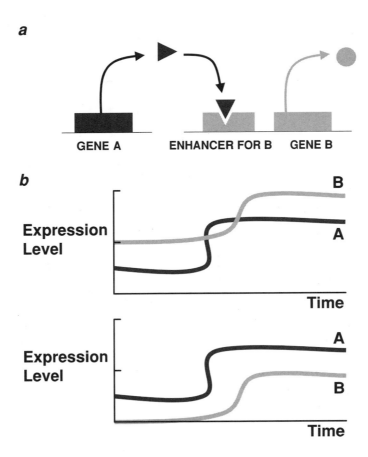

Figure 3.26
Dynamic relationships between genes. (*a*) The expressed product of Gene A binds an enhancer
region that increases transcription of Gene B. (*b*) Gene B's initial expression level before being
affected by Gene A can vary throughout the experiment. As a result, measuring the correlation
between the absolute levels of Genes A and B will not reveal the underlying enhancement
relationship between the two. Instead, this can only be done by analyzing the expression
dynamics—the change in expression level of Gene B in relation to the expression level of Gene
A. (Derived from Reis *et al.* [150].)

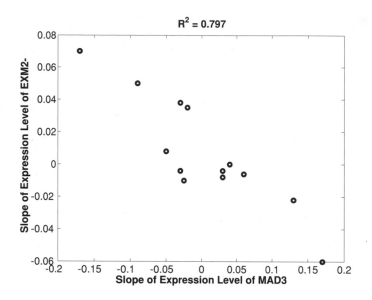

Figure 3.27
Negative dynamic correlation. The distribution of slopes of MAD3 and EXM2 plotted one
against another.

dynamics analysis. However, only about half of the 133 shared genes appear linked
to the same genes in both analyses—most appear linked to other genes.

Of many examples, figure 3.27 shows the distribution of slopes between RAD6
and MET18 which are highly correlated by the slope similarity measure but not by
the static gene expression levels. RAD6 is a ubiquitin-conjugating enzyme concentrated in the nucleus that is essential to mediating the degradation of amino end
rule-dependent protein substrates(21). MET18, also known as MMS19, is a protein
concentrated in the nucleus that affects RNA polymerase II transcription[150]).
These are inversely related in their dynamics, with an R^2 of -0.791. It is not
surprising that a gene responsible for protein degradation might have an inverse
relationship to a gene responsible for RNA transcription leading to protein synthesis. Again, the significance here is that the discovery of this relationship depended
upon picking the right similarity measure. Besides this one, there is undoubtedly
a large unexplored universe of potentially useful measures.

- *Diseases:* Tissues from single or multiple diseases can be clustered. What is currently thought to be one disease may in fact be distinguishable into two or more subtypes.

- *Phenotype:* Cell lines, organisms, or patients can be clustered based on a series of measurements on them, of which gene expression may be only one component.

- *Patients:* In medical informatics, patient clusters are distinguishable in many ways: geographical location or home, presence or absence of a disease, the quantitation of specific laboratory measurements, or details of a longitudinal course of medical care.

- *Promoters:* Transcription factors are known to bind these specific regions of DNA to cause increased or decreased expression of the controlled gene. These could be clustered based on sequence similarity or association with functional genomics measurements.

- *Environment:* Environmental toxins and factors can be clustered by the pattern of gene expression, or the change in expression, seen after exposure.

As diagrammed in figure 2.11, all of these data types may end up in a large heterogeneous data table that can be used to advantage in the automated data-mining techniques described below. It is precisely the lack of *a priori* limitations or assumptions of which data sources and data types will be of clinical or biological interest that is one of the bulwarks of the genomic approach to discovery.

4.3   What Does it Mean to Cluster?

There are certainly laboratories that criticize data mining or clustering as a "fishing expedition," believing that the analytic technique starts with absolutely no hypothesis. There are also laboratories that apply every existing clustering method to their microarray data sets, hoping to find some "significant genes or clusters." The ideal analytic method lies between these two approaches. Certainly, different clustering methods are useful for different purposes. One should not blindly apply every clustering algorithm without knowing which method is best for answering particular kinds of questions. The analyst should start the discovery process with a hypothesis; this hypothesis may not be limited to a particular gene or set of genes but should ask a specific question of the data. At this point, it may be useful to list some potential questions one could start with:

4 Genomic Data-Mining Techniques

4.1 Introduction

This chapter starts with an analysis of the components in molecular biology for which analysis can be performed. We then move to defining genomic data mining and what it actually means biologically to apply these clustering techniques in this domain. Although many analytic techniques are currently available, one should clearly not apply these techniques blindly without some hypothesis, so we discuss potential broad hypotheses one can start with. We then cover techniques for data reduction and filtering. Using examples written in pseudo-code, we cover the most commonly used techniques, discussing the advantages and disadvantages of each technique. We will then address the postanalysis process, including determining significance through permutation testing. Finally, we end with some thoughts on the automatic determination of genetic networks, including bayesian networks.

4.2 What Can Be Clustered in Functional Genomics?

Before launching into a discussion of the various clustering algorithms available to bioinformaticians, one needs to carefully consider the motivations behind data mining. *Data mining* can be loosely defined as determining the relationships of the elements in a data set. A variety of objects in biology can be captured in functional genomics data sets, a few of which should be mentioned here.

- *Genes:* Certainly, the majority of functional genomics bioinformatics analysis involves determining the relationships among genes. Genes can be clustered not only by gene expression but also by gene sequence, nucleotide composition, linkage, and chromosomal position. Genes can obviously be clustered using measurements from normal physiological states, as well as abnormal or pathological states.

- *Alternative splicing products:* An increasing number of genes are now predicted to have alternative splicing products, or a varying combination of exons. Various alternatively spliced products could be clustered to assess their pattern of expression compared to each other.

- *Tissues:* Normal tissues can be clustered to assess their relative degrees of similarity or dissimilarity. Unknown tissues, such as pathological specimens, could then be compared with clusters of known tissues.

- What uncategorized genes have an expression pattern similar to these genes that are well characterized?

- How different is the pattern of expression of gene X from other genes?

- What genes closely share a pattern of expression with gene X?

- What category of function might gene X belong to?

- What are all the pairs of genes that closely share patterns of expression?

- Are there subtypes of disease X discernible by tissue gene expression?

- What tissue is this sample tissue closest to?

- Which are the different patterns of gene expression?

- Which genes have a pattern that may have been a result of the influence of gene X?

- Which genes have a pattern that caused the expression pattern of gene X?

- What are all the gene-gene interactions present among these tissue samples?

- Which genes best differentiate these two group of tissues?

- Which gene-gene interactions best differentiate these two groups of tissue samples?

As we shall see as we progress through this chapter, different algorithms are more particularly suited to answer some of these hypotheses, compared with others.

4.4   Hierarchy of Bioinformatics Algorithms Available in Functional Genomics

Although there are only a few commonly used analytic techniques in functional genomics, it is important to remember that these techniques exist in a hierarchy of methods. One can model this taxonomy of techniques many different ways, as outlined in section 2.2; in this section, we chose to distinguish these methods as *supervised* or *unsupervised* [73]. An example of supervised learning is simply calculating the fold increase or decrease in expression of each gene in two microarray measurements. In this type of analysis, the particular context of each measurement is known, and the end result is a list of individual genes behaving differently in each of the different contexts. Another example of supervised learning is when it is algorithmically determined which genes are best able to determine an external labeling,[1] such as distinguishing one type of leukemia from another [78]. This type of approach is used in neural network construction [155], and decision tree construction [146], which can then be easily translated into diagnostic algorithms [33, 98].

In unsupervised learning, the patterns inherent in the data itself are determined, without *a priori* labels or contexts. Relevance networks, dendrograms, and self-organizing maps analyze every possible pair of genes to determine whether a functional relationship exists between them. This is determined using a metric, such as Euclidean distance between the genes (when genes are positioned in a multi-dimensional space with each experiment as a separate axis), Pearson's correlation coefficient (assumes a linear model of control between the two genes), or mutual information (with no assumptions as to the model of control). The end result of this analysis is a ranked list of hypotheses of pairs of genes that work together. A taxonomy of techniques divided in this manner would look like this:

1. **Unsupervised**: Analysis looking for characterization of the components of the data set, without *a priori* input on cases or genes.

   (a) ***Feature determination***: Determine genes with interesting properties, without specifically looking for a particular pattern determined *a priori*.

      i. Principal components analysis: Determine genes explaining the majority of the variance in the data set.

   (b) ***Cluster determination***: Determine groups of genes or samples with similar patterns of gene expression.

      i. Nearest neighbor clustering: The number of clusters is decided first, the clusters are calculated, then each gene is assigned to a single cluster.

         A. Self-organizing maps: Groups are found by iteratively moving each cluster centroid toward a randomly chosen data point; the center of each cluster is a hypothetical gene expression pattern.

         B. K-means clustering: Similar to self-organizing maps, except there is no map structure for the centroids.

      ii. Agglomerative clustering: Bottom up method, where clusters start as empty, then genes are successively added to the existing clusters.

         A. Dendrograms: Groups are defined as sub-trees in a phylogenetic-type tree created by a comprehensive pairwise dissimilarity measure, such as the correlation coefficient.

---

[1] For example, human expert-derived labels.

       B. Two-dimensional dendrograms: Both genes and samples are clustered separately.

  iii. Divisive or partitional clustering: Top-down method, where large clusters are successively broken into smaller ones, until each subcluster contains only one gene.

       A. Dendrograms.

       B. Two-dimensional dendrograms.

(c) ***Network determination***: Determine graphs representing gene-gene or gene-phenotype interactions.

  i. Boolean networks: Gene expression measurements are binarized; nodes in the graph represent an entire expression state, and links represent transitions from one expression state to another.

  ii. Bayesian networks: Nodes in the graph represent features (genes or phenotypic measures), and links between features are determined probabilistically.

  iii. Relevance networks: Nodes in the graph represent features (genes or phenotypic measures); links between features are calculated with pairwise application of a dissimilarity measurement, and networks represent agglomerations of these pairwise associations.

2. **Supervised**: Analysis to determine ways to accurately split into or predict groups of samples or diseases.

(a) ***Single feature or sample determination***: Find genes or samples that match a particular *a priori* pattern

  i. Nearest neighbor: Find genes or samples closest to another pattern, using a dissimilarity measure, such as Euclidean distance or the correlation coefficient.

       A. Comparison to hypothetical pattern: Find genes or samples closest in similarity to a hypothetical pattern, determined with *a priori* information.

       B. Comparison to actual pattern: Find genes or samples closest in similarity to one of the genes or samples in the data set.

       C. Comparison to clusters: First, remember the training set of data and labels; then, when presented with a test gene or sample, find the element of the training set that is closest and use its label.

  ii. Student's *t*-test: Find genes where the expression measurements are statistically different between groups of samples

(b) ***Multiple-feature determination***: Find combinations of genes (*i.e.*, more than one) that match a particular *a priori* pattern.

i. Decision trees: Use the training set of genes or samples to construct a decision tree to help classify test samples or test genes.

ii. Support vector machines: First take the set of measured genes, then create a richer feature set with combinations of genes, then find a hyperplane that linearly separates groups of samples in this larger multidimensional space.

The hierarchy above is meant to serve as a way to categorize bioinformatics techniques, and the examples under each branch of this hierarchy are incomplete; newer methodologies in each branch are shown in section 4.10.

This hierarchy just touches upon one of the important goals of the bioinformatics arm of functional genomics. That is, to ascertain gene regulatory networks from gene expression data sets. Few of the algorithms listed above find their way into this endeavor for reasons that we address here.

One crude definition of a regulatory network is a set of genes G and a set of relations R that relates any two elements from G. The relation may specify either a positive or a negative interaction. For example $(A, B, +)$ may indicate gene A interacts with gene B in a positive manner, such that the level of expression of gene B increases as the level of expression of gene A increases. Completely missing are alternative models other than these simple linear models of gene-gene interactions. Any element of translation into protein, the actual actions of proteins in DNA binding, any sense of temporality, and gene and protein degradation are missing. However, despite all that is missing, there is enough captured in this simple model to provide material to construct a biological hypothesis linking the two genes. [177]

In the above hierarchy, we do not consider a dendrogram to be a member of the network determination category of algorithms. Why is this? One typically constructs dendrograms from a gene expression data set to survey the types of expression patterns, and to determine the neighbors of genes to be classified. Although dendrograms are computed from a comprehensive pairwise calculation of gene-gene expression pattern similarity, as detailed below, these gene-gene links are not shown in the output. Besides the first gene-gene pair, each subsequent gene is added to the dendrogram when its expression pattern matches an expression pattern of a branch. These branch expression patterns represent the average expression pattern of the descendant genes, and may not match any one of these genes. Furthermore, gene-gene interactions with negative correlations are not represented in dendrograms, thereby eliminating a large class of biological regulatory behavior.

In the same vein, why do we not consider a self-organizing map as part of the network determination category of algorithms? Self-organizing maps are typically made to survey the types of expression patterns seen in a gene expression data set. Like dendrograms, self-organizing maps do not display gene-gene interactions; instead, these typically output a centroid expression pattern (essentially, an average expression pattern not matching any one of the genes in the data set) and output the expression patterns of the genes surrounding the centroid. There is no representation of a gene-gene interaction, nor are negative interactions considered.

4.5 Data Reduction and Filtering

Clustering algorithms can easily be biased if certain assumptions they hold are not met. Each dissimilarity measure has its own set of assumptions and requirements. For example, the reliability of the correlation coefficient not only assumes that two data sets are normally distributed but that they are easily biased by outlying points (see figure 4.1). Another example is that genes with expression measurements that are constant across samples may still show variation due to measurement noise, and if these genes are not filtered out, such measurement noise can be amplified by normalization and can appear as a true signal. Four common methods for filtering genes meeting these degenerate cases are given here.

4.5.1 Variation filter

One way to prevent degenerate clustering is to remove genes or samples that are changing "too little." A variation filter is used to ensure that genes or samples have a wide enough dynamic range of measurements. Dynamism can be measured in both absolute and relative terms. For example, a gene could be removed from further analysis if there is less than a twofold difference in the measurements across the samples (a relative term) or if there is less than 35 expression units of difference in the measurements across the samples (an absolute term).

```
make a new empty list called post_filter
loop through each gene G to be filtered
       find the minimum expression level for G
       find the maximum expression level for G
       set the flag keep_gene to false
       if maximum - minimum is over 35
           set keep_gene to true
       if maximum / minimum is over 2
           set keep_gene to true
       if the flag keep_gene is true
           add this gene G to the post_filter list
```

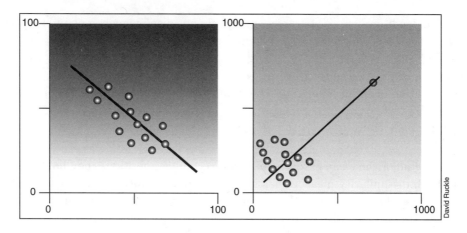

Figure 4.1
Example of how a single point can distort overall correlation. On the left: the scatterplot shows
a negative correlation. The scatterplot on the right is identical, save for the additional point
added. If the values of even a single point are high enough, the correlation coefficient can be
altered (though the variance of the correlation coefficient, if calculated, would be higher). This is
primarily because using the correlation coefficient assumes values distributed normally.

```
end loop
```

There are those investigators who will object to this kind of filter because there
may well be biologically important changes in expression that are very small and will
be removed by this variation filter. However, if functional genomics experiments are
seen as a funnel (see figure 1.6) for determining which are the high-yield hypotheses
to pursue for biological testing, then the high false-positive rate of genes hypothesized wrongly to be biologically relevant may prove to be costly (see figure 2.9 in
section 2.1.4).

4.5.2   Low entropy filter

The opposite problem can also occur with gene expression patterns. Genes can
demonstrate spiking behavior, where low expression levels are seen in all samples
except one. The single high expression can dominate a pairwise analysis using
correlation coefficients, for example (see figure 4.1 for an example) [94].

An entropy filter can be used to remove genes that demonstrate spiking behavior,
or, in other words, that are not well distributed over its range of values. Entropy is
a measure of the amount of disorder in a variable. If measurements of a variable are
equally distributed across its dynamic range, then very little information is known

about the expected value for the variable. Once the measurement is known, the amount of information about the variable increases greatly, meaning such a variable has high entropy. Compare this to a variable that can be measured at only two values. A great deal of *a priori* information is already known about the expected value for this variable; thus, once the measurement is actually known, the amount of information increase is less. This indicates the variable has lower entropy (see figure 4.2).

$$H(x) = -\sum_i p(x_i) \log_2(p(x_i))$$

where x is the variable whose entropy H is being calculated, $\log_2$ is base 2 logarithm, and $p(x_i)$ is the probability a value[2] of x was within quantile i of that feature. For example, if one were using 10 quantiles, and a gene with the expression amounts 20, 22, 60, 80 and 90 would have deciles 7 units wide, with two values in the first decile, one in the sixth decile, and one in the ninth and tenth decile, making $H = -1.92$. This pseudocode accomplishes the same.

```
define number_bins as the number of bins to use in entropy
    calculations
make a new empty list called post_filter
loop through each gene G to be filtered
    find the minimum expression level for G
    find the maximum expression level for G
    set expression_range to maximum - minimum
    set interval to expression_range divided by the number_bins
    make a new empty array called deciles, with size = number_bins
    set deciles[1] to interval
    loop through the rest of deciles[2..number_bins] using the
        index i
        set deciles[i] to deciles[i-1] plus interval
    end loop
    make a new empty array called decile_count, with
        size = number_bins
    loop through all the expression measurements for G
        set the flag bin_found to false
```

[2] An expression value if the variable whose entropy is being calculated is a measure of gene expression, but any kind of variable, even a noncontinuous or categorical variable such as eye color, can be used in an entropy calculation.

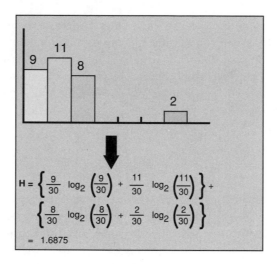

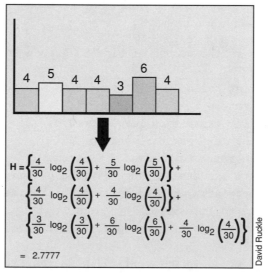

Figure 4.2
Detecting spikes by calculating entropy. The top graph shows a hypothetical gene with
"spiking" behavior; *i.e.*, the gene expression is markedly higher in two samples, compared to the
other samples. The bottom graph shows a second hypothetical gene with gene expression
measurements that are more distributed across the dynamic range. Distribution of the gene
expression measurements are not just a characteristic of the gene; it is also connected to the
samples in which expression was measured.

```
       loop through deciles[1..number_bins] using the index i
           if the expression measurement is <=
               deciles[i] and not bin_found
           add one to decile_count[i]
           set the flag bin_found to true
       end loop
   end loop
   set entropy to zero
   loop through decile_count[1..number_bins] using the index i
       if decile_count[i] is over zero
           set probability_in_bin to decile_count[i] divided by
               the number of expression measurements in G
           add to entropy (probability_in_bin times
               log-base-2( probability_in_bin ) )
       end if
   end loop
   set entropy to -1 times entropy
end loop
```

After entropy is calculated for each gene or sample, a threshold low entropy (*e.g.*, at the lower 5th percentile) can be chosen and used to filter out genes that do not display a sufficient range of values. In using this technique, however, one needs to note that the calculation of entropy does directly depend on the number of quantiles used, and the use of 10 deciles is arbitrary.

As an alternative method, Heyer *et al.* [94] proposed the use of the jackknife correlation coefficient to counter the spiking problem. The jackknife correlation coefficient is an alternative dissimilarity measure to the standard (Pearson's) correlation coefficient. To compute this measure for two genes measured in n samples, the technique involves computing n different correlation coefficients, each time with one of the samples removed. The jackknife correlation coefficient is then the minimum of the separate correlation coefficients. Use of this dissimilarity measure will effectively remove single outliers. However, it is not clear whether removing a single outlier is enough, especially when samples from many disparate tissues are used, and several spikes are seen. It is also questionable whether removing spikes is even desirable, because when two genes demonstrate spiking behavior in the same samples, it may reflect something biologically interesting.

It is important to note that both variation filters and low entropy filters remove entire genes from downstream analysis. If the subsequent analysis is being used for hypothesis generation, this means that hypotheses containing the filtered genes will not be created or considered. Yet again, the cost-benefit analysis of section 2.1.4 should be considered.

4.5.3   Minimum expression level filter

As described elsewhere in this book, it is commonly believed that low gene expression measurements are less reliable than high gene expression measurements. Empirically, this means one's confidence interval around a measurement differs as a function of the expression level. If the low expression levels are particularly noisy, this can cause artifacts in the downstream clustering process, or worse, cause the creation of incorrect clusters. One solution to this is to ignore those gene expression measurements with a measurement under a threshold value. Threshold minimum expression values can be calculated arbitrarily at a percentile of the distribution of expression measurements (*e.g.*, at the lower 5th percentile), determined empirically at an ideal point for clustering sensitivity and specificity [40]. Here, we show pseudocode that removes genes using the first of these three methods.

```
loop through each gene G to be filtered
    find the minimum expression level for G
    find the maximum expression level for G
    set expression_range to maximum - minimum
    set lowest_5th_percentile to expression_range times 0.05
    loop through all the expression measurements for G
      if the expression measurement is under lowest_5th_percentile
        remove that measurement from G
    end loop
end loop
```

Note that with this type of filter, we are removing gene expression measurements, not entire genes, from the downstream analysis. For example, when genes A and B are compared using a dissimilarity measure, each sample normally has two measurements associated with it: the expression level of gene A in that sample, and the expression level of gene B in that sample. By removing gene expression measurements using a minimum expression level filter, we are removing one or both of those gene expression measurements from the sample. Thus, when genes A and B are compared using the dissimilarity measure, there will be fewer samples used in calculating the measure.

4.5.4   Target ambiguity filter

Another seemingly straightforward filter has not been used as commonly. Given the accelerating rate of new annotations in DNA databases, it is not uncommon that a probe or probe set used on an older microarray is no longer specific for a single gene. This can happen if the probe was either designed without knowledge of the entire genome or designed against a piece of chromosome instead of an expressed sequence. As an example, probes made against GenBank accession AF053356 might represent the expression of any of 12 expressed products, since this sequence in GenBank represents a piece of chromosome 7. At the time of the microarray construction, however, this region may have been thought to represent only a single gene.

Since without knowledge of the exact probe sequences it is not clear what an expression measurement on a probe set made from a piece of chromosome means, one can argue for removal of these types of probe sets from analysis at the start. However, this requires the use of probe databases early in the analysis, which is not easily performed.

4.6   Self-Organizing Maps

The use of self-organizing maps (SOM) in functional genomics was first described by Tamayo *et al.* [175]. The underlying hypothesis in this technique is the standard functional genomics dogma articulated in section 2.2: genes with similar expression patterns are functionally similar. Thus, if one can find groups of genes that are all near each other, such a group is likely to be related functionally.

As typically used, this methodology starts by rescaling each gene into a standard distribution, *e.g.*, with mean = 0 and variance = 1, though this is not required. Each measured gene is mapped into a k-dimensional point, such that each of the k dimensions corresponds to a different tissue sample or microarray. The ith coordinate axis represents the ith sample. Each gene is evaluated by its vector of measurements[3] and is represented by a point in the map. Coordinates for these points therefore represent the expression levels in various experiments. This is easy to visualize for a gene measured in three experiments (see figure 4.3). The multidimensional space would be represented as a hypercube. Each side of the cube represents a different tissue sample or microarray. Each gene would be a multidimensional point within the cube space, with each coordinate equal to its measurement.

---

[3]Each measurement of that gene in each experiment corresponds to one element of that vector.

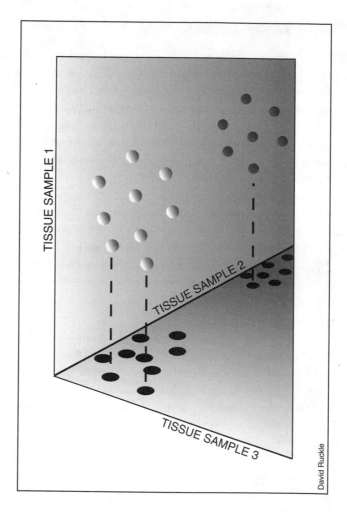

Figure 4.3
Genes can be represented as points in a multidimensional space. Each sphere represents a gene measured in three tissue samples. Each of the expression levels is considered as a measure along a three-dimensional coordinate. One can imagine that genes with similar expression levels in all three samples may be grouped as clusters in the three-dimensional space. These clusters can be found using an unsupervised technique, such as self-organizing maps.

The analysis starts by choosing a map or topology of nodes N (called centroids), typically arranged in a grid.[4] Each node is in the same dimensional space as the genes, or objects being clustered. Each centroid starts at a random position in the k-dimensional space as follows:

```
define number_centroids as the number of centroids we will start with
define number_axes as the number of samples or experiments for which
     we have data
define vector as an array of floating point numbers,
     with size = number_axes
make a new empty array of vectors called centroids,
     with size = number_centroids
loop through centroids using the index i
     loop from 1 to number_axes, using index j
          find the minimum expression level across all the
               genes measured in sample j
          find the maximum expression level across all the
               genes measured in sample j
          set coordinate j for centroid i to be a random number
               between the minimum and maximum
     end loop
end loop
```

The centroids are subsequently mapped to a position in this k-dimensional space and correspond to the center of a cluster. The mapping process is iterative, meaning that the map is formed over several cycles. During each cycle, a random data point is chosen and all of the centroids are moved toward that point, as illustrated in figure 4.4. The distance each centroid is moved is typically a function of both the iteration number and the distance between the point and centroid, such that the distance moved decreases as more iterations have passed. The distance moved is also less for those centroids farther away from the randomly chosen point.

```
define number_centroids as the number of centroids we will start with
define number_genes as the number of genes we are working with
set continue_loop to true
set iteration to zero
```

---

[4]Note that this topology is selected by the investigator sometimes on the basis of prior knowledge (the number of expected natural clusters of the data) and sometimes empirically by iterating across several possible topologies and inspecting the results.

```
loop while continue_loop
    set cumulative_moving to a zero vector
    calculate learning_rate, as a function of iteration
    pick a random gene G from the list of all genes we have
    loop through centroids using the index c
        calculate dissimilarity measure (i.e. distance) from
            centroid to G
        calculate vector v by taking centroid c and subtracting G
        move centroid c along vector v as a function of
            learning_rate
        add the amount moved to cumulative_moving
    end loop
    add one to iteration
    set average_moved to cumulative_moving divided by
        number_centroids
    if average_moved is under a predetermined threshold
        set continue_loop to false
    end if
end loop
```

The algorithm continues until either fixed endpoints are reached, or after there is no movement of nodes above a threshold.

As most commonly used, self-organizing maps employ Euclidean distance as the dissimilarity measure. However, there is nothing that would prevent the use of the correlation coefficient or mutual information as the dissimilarity measure instead. Nonetheless, we describe this algorithm using the Euclidean distance because that is the way it is typically presented in the most well-known applications in functional genomics [78].

Figure 4.5 explains the conventional graphical representation of self-organizing maps.

4.6.1   *K*-means clustering

The k-means clustering algorithm is very similar to the self-organizing map methodology. Centroids are initialized to randomly chosen genes, then each gene is mapped to its closest centroid. In an iterative manner, the centroids are then updated with the average expression pattern of the genes assigned to that cluster; then the genes are again mapped to their closest centroid. This continues until all genes remain mapped to the same centroids.

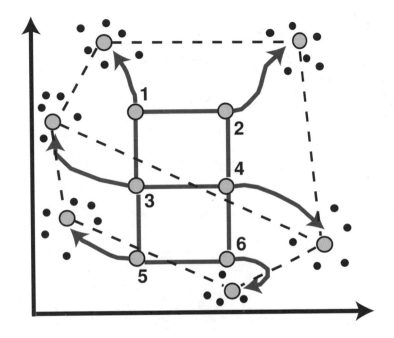

Figure 4.4
Principle of self-organized maps. Centroids start in an arbitrary topology; then, as the method progresses, each moves toward a randomly chosen gene during each iteration. After proceeding for enough time, each centroid will be in the middle of a cluster. (From Tamayo *et al.* [175].)

Each cell in the grid represents a
separate cluster. Typically, genes can
belong to only one cluster.

Cluster number (can be arbitrary)

Representative (or
average) expression
level pattern from
genes in cluster.

Grid represents
initial topology of
centroids (in this
case, a 5 by 6 grid)

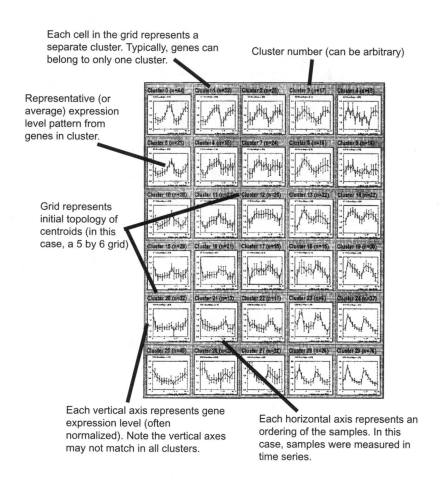

Each vertical axis represents gene
expression level (often
normalized). Note the vertical axes
may not match in all clusters.

Each horizontal axis represents an
ordering of the samples. In this
case, samples were measured in
time series.

Figure 4.5
Graphical convention for self-organizing maps. (Derived from Tamayo *et al.* [175].)

Published examples

- Tamayo *et al.* [175] used self-organizing maps to functionally cluster genes into various patterned time courses in HL-60 cell macrophage differentiation.

- Toronen *et al.* [180] used a hierarchical self-organizing map to cluster yeast genes responsible for diauxic shift.

- Soukas *et al.* [165] used a k-means approach to cluster genes involved in leptin signaling.

- Tavazoie *et al.* [176] used k-means to determine clusters of genes based on function, for which $5'$ upstream regions in the genome were searched in an effort to find common regulatory sequences.

Advantages

- Easy visualization in two dimensions of the range of variation seen in the data: Since the map will be spread across the multidimensional space of experiments, by displaying the expression patterns represented at each node, one can quickly see the variety of gene expression patterns. For example, if there was not much variety in the gene expression patterns, one would quickly find the representative gene expression patterns from the many nodes of the self-organizing map appearing the same.

- Computational requirements: Since only the centroids are moving in each iteration, the number of distances to be computed is linear with respect to the number of nodes. After the map is made, gene assignment to a cluster may require the computation of distances from each gene to each centroid. On the face of things, this is still much less computationally expensive than the initial precomputation of comprehensive pairwise dissimilarity measures required by the other techniques, such as dendrograms and relevance networks. However, this gain in performance is tempered by the large (but finite) number of iterations that may be required to reach stability.

Disadvantages

- Arbitrary map topology: There is nothing to indicate exactly how to create the initial map geometry. The k-means algorithm requires the number of nodes to be specified as *a priori* information. The self-organizing map technique also requires the initial geometry of the nodes to be specified.

- Misses negative correlations: Negative correlations include gene interactions such as those demonstrated by tumor suppressor genes. As an example, *p53* suppresses the expression of several other genes. This means that the higher the expression of *p53* seen, the less expression of other genes is expected. Negative correlations will be missed in this type of approach.

- Influenced by spiking in gene expression: Without adequate filtering, spiking (isolated high or low expression in single samples) can prevent proper normalization, which impairs self-organizing map formation.

- Each experiment or coordinate is treated independently and with the same weight: each axis is treated as orthogonal to the others. This may not be the actual case, however. Consider the extreme example of a situation where two replicated measurements are made on three patients, making six microarray measurements. In each of the three patients, the pair of replicated measurements should not differ. However, if each of the six measurements is treated as a separate, orthogonal axis, too much weight may be given to the replicates, which may have measurements close together. In other words, genes may appear closer simply because they have many similar measurements across the replicated experiments.

- Further analysis is needed to determine the actual boundaries of each cluster: After the map is made, all the genes in the multidimensional space may be assigned to a cluster by choosing the centroid closest to each gene. However, it is not immediately clear where the boundaries of each cluster lie. First, this is important in determining which genes lie within the cluster. Second, this is crucial because it is important to distinguish those genes that are far away from all centroids. If too many genes are too distant from all the clusters, it can be an indication that additional centroids need to be added at the at the start of analysis to gain coverage in that area of the multidimensional space. Thus, it is important to determine what the shape and size of the clusters are, but this requires prior assumptions (the space of the centroids can be round or oval, the diameter can be large or small, or equal among all the clusters, *etc.*) for which there may be little domain or expert knowledge.

- Membership of a gene may be limited to a single cluster: If a gene resembles the centroid pattern seen in two clusters, it may be assigned to one of the two clusters but not both, unless precautions are taken to scan for this situation.

- Missing data: If a single gene expression measurement is missing, it becomes difficult to determine where to place that gene in the multidimensional space. This positioning may be crucial to determining which centroid is closest to that gene.

- Stochastic technique: Since the initial conditions are determined randomly, the results may not be repeatable. Validation techniques can be performed by repeating the algorithm using several random starting positions, but it is not clear how to merge the resulting, differing maps.

4.7 Finding Genes That Split Sets

As suggested by the machine-learning taxonomy above (section 4.4), nearest neighbor analysis has been used in both supervised and unsupervised learning methods. In a supervised application, the technique is traditionally used to find genes that match an externally specified pattern. These patterns may be "ground truths" in independently validated biological knowledge (*e.g.*, a gene known to be involved in the process being studied) or empirical (*e.g.*, a gene that is observed to be highly expressed in certain samples and expressed at a lower level in other samples). In an unsupervised manner, the technique is used to find clusters of genes that share similar expression patterns.

Let us start with the supervised method. We may have a specified gene of interest in our data sets. Alternatively, we may specify a desired or hypothetical gene expression pattern which may not exist in the data set. Such a pattern could be made based on the desired properties of a gene in particular samples or patients. For example, one may want to find a gene that was upregulated in patients with disease and downregulated in nondiseased patients.

To get a list of genes related to either query pattern, we must again define what it means to have similar patterns. Although any dissimilarity measure can be used as the judge of whether expression patterns are similar, Euclidean distance and the correlation coefficient have been used traditionally. Iteration through the data set will quickly find and rank genes by the degree of similarity to the query pattern.

```
define number_genes as the number of genes we are working with
define query_pattern Q as a vector across all the samples,
      representing the gene with the expression pattern we are most
      interested in
make a new empty array called distances, with size = number_genes
loop through all the genes, with index G
      calculate dissimilarity measure (i.e. distance) between Q to G
      store the measure in the appropriate position in distances
end loop
```

Published examples

- Using a hypothetical pattern, Golub *et al.* [78] used this technique to find those genes that best split samples of acute lymphocytic leukemia from acute myelogenous leukemia.

- Ben-Dor *et al.* [19] used a nearest neighbor classifier to split normal colon samples from cancer samples from [4], as well as other data sets.

Advantages

- Computational and memory requirements: When used in a supervised manner, nearest neighbor analysis is considered a "lazy classifier," in that the algorithm requires only that the data set be kept in memory. The genes "nearest" to the search pattern are found when required, minimizing unnecessary comparison operations. This is in contraste to other techniques, which may require the initial expensive computation of pairwise dissimilarity measures.

- Quickly finds genes or features that most significantly split the labeled sets. These can be validated biologically, then developed into diagnostic tools, for example.

Disadvantages

- Only gross differences are found. In other words, genes with differences in absolute expression level are typically chosen. Differences in gene-gene interactions may be missed where the expression level of the genes involved in the interaction may not be different (see figure 4.6).

- The genes or features that best split two sets may not necessarily be the most significant or biologically causative. For instance, the genes selected may be the most obvious but distant "downstream" effects of other genes which are primarily responsible for the difference in states.

- Because many more features (*i.e.*, genes) are measured compared to cases (*i.e.*, samples or experiments), it is almost always possible to find genes that split samples into labeled sets. If one considers combinations of genes as features (*i.e.*, with support vector machines), then one can find even more features that successfully split the sets. One has to be careful to look for simple models that split sets, even if they are not as accurate [30]. Otherwise, these models may lose biological relevance.

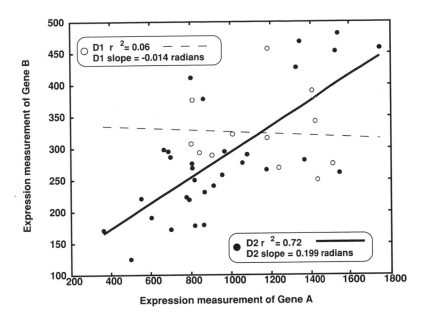

Figure 4.6
Genes can have a difference in interaction, but not in expression level. Scatterplot of gene A and gene B, measured in samples from disease 1 (open circles) and disease 2 (closed circles). Note that expression measurements from neither gene A nor gene B can be used to separate disease 1 from disease 2. However, the linear regression model of gene expression levels from disease 2 is different from disease 1.

4.8 Phylogenetic-Type Trees

A third methodology is phylogenetic-type[5] tree construction, otherwise known as *dendrograms*. Although there are several ways to construct this data structure, this technique typically first involves comprehensively comparing all genes against each other using a dissimilarity measure, such as the correlation coefficient. Eisen *et al.* [63] took expression levels at various time points and created a vector for each gene. They then compared all genes against each other and recorded the correlation coefficient for each pair of vectors, then constructed a phylogenetic-type tree with branch lengths proportional to the correlation coefficients.

Let us take this process one step at a time. First, enough memory must be allocated to store the comprehensive pairwise comparisons. If n genes are being compared with each other, there are $((n*n)/2) - n/2$ possible pairwise comparisons. Typically, when maximum efficiency is used, dissimilarity measures are stored as real values taking 4 bytes each. Clearly, as n increases, the necessary memory increases on the order of n^2. Next, an iterative method must be used to perform the pairwise comparisons. Since associativity holds with the correlation coefficient and Euclidean distance, once gene A is compared with gene B, the reverse comparison does not need to be performed. There are four logical ways to order these comparisons, as shown in figure 4.7, and one needs to be chosen arbitrarily. Regardless of how we are calculating or storing the measures in this chapter, we refer to the comprehensive set of measures for a collection of variables or features as a *measure-triangle*.

Although there are many efficient ways to store the results of each comparison, *e.g.*, using matrices, we purposely use a less efficient, but easier-to-understand method here. We store the result of each comparison in a hash table, using the two genes (sorted) as the key. That is, the dissimilarity measure can be retrieved from the hash table using the unique pair of genes.

```
define number_genes as the number of genes we are working with
make a new empty two-dimensional array called measure-triangle,
     with size = number_genes by number_genes
loop from 2 to number_genes, with index i
     loop from 1 to i - 1, with index j
          calculate dissimilarity measure (i.e., correlation
               coefficient) between gene[i] and gene[j]
```

---

[5]They do not describe the distance between the genomes of organisms as in standard phylogenetic trees and therefore are referred to as "phylogenetic-type" trees.

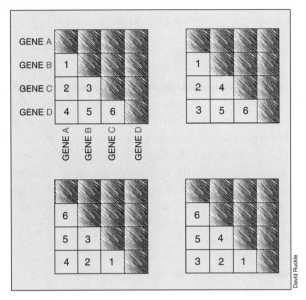

Figure 4.7
Four possible ways to order comprehensive pairwise comparisons. We define the term
measure-triangle as the comprehensive pairwise comparison between each of a set of features.
The comparisons can be performed and stored in any of four different ways.

```
        store measure in measure-triangle[i][j]
    end loop
end loop
```

For each possible pair, the dissimilarity measure must be applied. Eisen *et al.* [63] essentially used the correlation coefficient in pairwise comparisons. Wen *et al.* [189] used Euclidean distance as a dissimilarity measure.

In addition to the measure-triangle used to store the dissimilarity measures, we need a data structure to hold the developing tree. A tree can be thought of as a set of internal nodes and external nodes (or leaves). Internal nodes serve as branch points linking together genes or subbranches. External nodes or leaves are exclusively genes. Because both genes and branches can be connected by branches, we will make that fact explicit by stating that they both can be branchable.

```
define number_genes as the number of genes
define number_samples as the number of samples
define an object called branchable, which contains
        (1) an array of expression measurements of size number_samples
        (2) a link called parent, to another object of type branchable
        (3) a list called children, to other objects of type branchable
define an object called branch-branch, which is derived from
        branchable
define an object called branch-gene, which is derived from branchable
make a new empty list called remaining-branchable-objects
make a new empty list called used-branchable-objects
loop through all the genes, with index g
        make an object of type branch-gene for each gene
        set the list of children for this object to null
        set the parent for this object to null
        add this object to the list remaining-branchable-objects
end loop
```

We are ready to begin building the tree. Iterations continue until every node has been attached to the tree. The tree begins with a single branch with two leaves, connecting the two genes with the strongest connection (*e.g.*, either the highest correlation coefficient or the shortest Euclidean distance). We will store the branches

```
loop until the size of remaining-branchable-objects reaches zero
```

```
    find the strongest measure in measure-triangle
    determine the two branchable objects, called a and b,
        that were compared to make this measure
    make a new object of type branch-branch, called c
    add the branchable objects a and b as children of c
    set the parent of branchable objects a and b to c
    set the expression measurements of c to the
        average expression measurements of a and b
    remove objects a and b from remaining-branchable-objects, if
        they were in that list
    add objects a and b to used-branchable-objects
    remove all comparisons involving a and b in measure-triangle
    compare object c to all the objects in remaining-branchable
        -objects and add those measures to measure-triangle
    if size of remaining-branchable-objects is zero
        set main-tree-trunk to c
    end if
end loop
```

The two genes are removed from the measure-triangle, or specifically, all the measures involving any one of these two genes are removed from the measure-triangle.[6] In place of the two genes, a new entity is created representing the branch with the two genes. The gene expression pattern for this new branch is the average pattern of the two gene expression patterns being linked. Notably, the new gene expression pattern for the branch does not have to be the average pattern. In computing the distance between a gene and a branch, or a branch and a branch, one could use a minimum, maximum, or average dissimilarity measure. New measures need to be calculated comparing the branch against all the remaining genes, and these are stored back in the measure-triangle. At every iteration, the next strongest measure is found and the process repeats, removing two entities (genes or branches) and replacing them with a new one, until only one entity remains.

At this point, the construction of the dendrogram is complete. However, the results still need to be displayed. The constructed dendrogram is typically used to order and display genes with similar clusters. At every branch, the display ordering can be one of two possibilities: Order one branch before the other, or *vice versa*. This means for any tree with n branch points, there are 2^{n-1} possible display orderings of the data. Other information needs to be used to determine

---

[6]That is, all the elements of the hash table for which one of the two genes is a key.

a consistent way to order the data. For example, position along a chromosome or average expression level can be used to order genes. Since there are no loops present in dendrograms, we can use a depth-first search to output the results. The following pseudocode will traverse the dendrogram and output the leaves of the tree (*i.e.*, the genes) in order by average expression level.

```
define a new procedure called print-tree, that takes a branchable
     object b as a parameter
if this branchable object b is a branch-gene (i.e., and not a
     branch-branch)
          print the name/label of b
end if
if the list of children of b is not empty
     move the left margin left 1 cm
     sort the list of children by expression level
     call procedure print-tree recursively on each child
          of b separately
     move the left margin back 1 cm
end if
end procedure


start the printing by calling print-tree on main-tree-trunk
```

The result is a dendrogram, typically drawn with the graphical conventions shown in figure 4.8.

4.8.1   Two-dimensional dendrograms

Using the technique above, one creates a dendrogram and the position of the branches suggests the presence of clusters. As mentioned above, the results still have to be displayed, and one can use a secondary key to order the data, such as average expression level, chromosomal position, *etc.*

We have not yet answered how to order the data horizontally; *i.e.*, how should the experiments be ordered, rather than the genes? For ordered experiments, such as time series data collection, gene expression is typically shown from earliest to latest acquisition. However, when the samples have no ordering, a second dendrogram is typically constructed to cluster the samples. This was the approach used by Ross *et al.* [154] in clustering both various samples of cancer cell lines and the genes measured by microarray, as shown in figure 4.9. The algorithm to construct such a

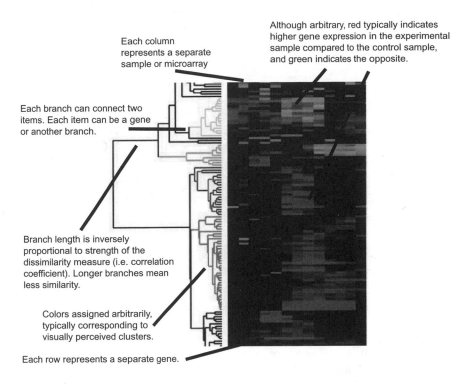

Figure 4.8
Graphical convention for dendrograms.

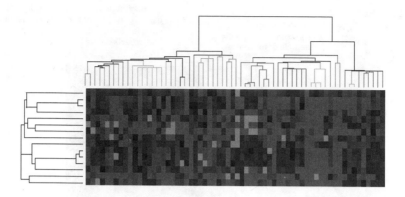

Figure 4.9
Two-dimensional dendrogram. Dendrograms are constructed for both the genes and samples individually, and are combined in the display. (Derived from Ross *et al.* [154].)

two-dimensional dendrogram is identical to constructing each of the dendrograms individually.

Published examples   There are many examples of dendrograms applied to gene expression measurements—too many to accurately list all of them. We will cite the earliest publications and the ones most cited.

- Spellman *et al.* [167] constructed dendrograms from a yeast expression data set constructed under multiple conditions.

- Eisen *et al.* [63] described the application of dendrograms to multiple yeast expression data sets, 1998.

- Alizadeh *et al.* [3] applied this technique to discover two subgroups within samples from the single disease B-cell lymphoma, and showed a statistically significant difference in survival curves of the patients from whom the samples were obtained.

- Ewing *et al.* [66] applied this technique after linking EST measurements with gene expression measurements.

- Bittner *et al.* [26] classified genes measured in malignant melanoma.

- Ross *et al.* [154] classified gene expression measurements from 60 cancer cell lines.

Advantages

- Category number and size: One can estimate the number or size of different expression patterns in the data set. For example, in viewing a self-organizing map, the user may not be sure there was enough coverage of the multidimensional space.

- Quickly visualize overall pattern similarity: One can quickly see whether most of the genes or samples in the data set are similar to or different from each other.

- Parallelizability: The most time-consuming aspect in the construction of dendrograms is the initial comprehensive precomputation of the dissimilarity measure across the set of features. This computation can be easily parallelized. The construction of the actual dendrogram requires the serial creation of branches, and is not trivial to parallelize.

- Display requirements: Optimally displaying dendrograms requires only a narrow column, in which gene names, symbols, and expression patterns (typically displayed in a color representation) can fit.

Disadvantages

- Misses negative correlations: Negative correlations include gene interactions such as those demonstrated by tumor suppressor genes. As an example, *p53* suppresses the expression of several other genes. This means that the higher the expression of *p53* seen, the less expression of other genes is expected. Negative correlations need to be explicitly searched for.

- Each gene is connected to the overall structure by only one link: the genes in a dendrogram are stored as leaves of the tree. Clearly, each leaf can only be connected to the tree by one stem. However, multiple links from a gene may be of interest. For example, one may wish to distinguish between a transcription factor responsible for the regulation of 10 genes and a structural protein whose gene is controlled by a single transcription factor. Dendrograms do not allow for a variable number of links from each gene.

- All genes are clustered, regardless of connection strength: the algorithm continues until every gene is connected. By definition, certain connections may be weaker than others. It may be a disadvantage to see all genes connected, when particular links are much weaker (*i.e.*, perhaps less likely to be validated) than others.

- Single structure: Although a dendrogram is a single structure, intuitively one can use a fixed distance from the root or a set of long branches to determine where to cut the branches in order to delineate the subgroups. This may or may not have a statistical basis.

- Single metric type: Dendrograms can only be constructed using a single dissimilarity measure. For example, it is not practical to construct a tree with genes connected by both Euclidean distance and correlation coefficient. Only one dissimilarity measure may be used at a time.

- Single data type: As stated at the start of this chapter, although we have many objects in functional genomics that may be clustered, it is only practical to cluster one data type at a time using dendrograms. For instance, genes may be clustered, and samples may be clustered. If both need to be clustered, this is performed as a two-dimensional construct using coupled two-way clustering [76]. Nonetheless, on occasion it does make sense to cluster genes and phenotypic measures along the same axis. A "systolic blood pressure" measurement might appear in one part of the dendrogram, while a "weight" may appear in a different part. It would be difficult to determine which of the genes in the immediate surrounding branches might be closest in profile to the phenotypic measure.

- It becomes difficult to follow the branching in a dendrogram, when the number of genes or features is high [93].

- As each new branch is added to the forming tree, the average gene profile representing a higher branch may not match any of the specific gene profiles within the subtree [162].

- These are not globally optimal trees: The algorithm described here falls under the category of "greedy algorithms" in computer science, as the best match is always used at every step, regardless of the later consequences it may have for forming the tree. Early "bad" decisions in the formation of branches cannot be corrected later.

- Entire vector is used: This means that for each gene, the entire set of samples are used in calculating the dissimilarity measurements. However, there may be very little variation in most of the samples, and high variation in only a few of the samples. Two genes might be scored as being similar, when they may actually differ greatly in one of the many samples.

- Computational requirements: This technique requires the comprehensive precomputation of the dissimilarity measure for all pairs of genes, which grows on the order of n^2 where n is the number of genes being studied. This can be a time-consuming computation.

4.9   Relevance Networks

Relevance networks offer a method for constructing networks of similarity, with the principal advantages being the ability to (1) include features of more than one data type, (2) represent multiple connections between features, and (3) capture negative as well as positive correlations. Like dendrogram construction, this algorithm begins by evaluating the similarity of features by comprehensively comparing all features with each other in a pairwise manner over the same cases. Several dissimilarity measures have been used in this methodology, including mutual information and the correlation coefficient. In this section, we describe the construction of relevance networks using $\hat{r}^2$. This is equivalent to r^2, except the sign (positive or negative) from the r, or Pearson's correlation coefficient, is applied. This is crucial in maintaining information about negative correlations. After the measure-triangle is completed, a threshold measure is chosen. The entire measure-triangle is scanned and any measures below the threshold are set to zero.

```
define threshold as the strongest link we are going to rule out
loop through all the measures in the measure-triangle
     if the measure is under threshold
          set the measure to zero in the measure-triangle
     else
          leave the measure alone
     end if
end loop
```

The measure-triangle must then be scanned to determine the networks. First, for the connections that remained because of measures that were greater than the threshold, each of the two nodes participating in the connection is listed as being used. An array stores the number of associations in which each gene is participating.

```
make a new empty array called node-connections, with
     size = number_genes
loop through all the measures in the measure-triangle
     if the measure is at or over threshold
          find the two genes A and B that were compared to make
               this measure
          add one to gene A's entry in the array node-connections
          add one to gene B's entry in the array node-connections
     end if
```

```
end loop
set number-nodes_used to zero
loop through the entries in the array node-connections
      if the entry is greater than zero, increment number-nodes-used
end loop
```

Next, each of the networks needs to be found and assigned a number. This is done using a combination linear-recursive methodology. A counter for the network number to be assigned is first set to 1. The list of features is iterated on, until a feature is found that is used in the relevance networks. That feature is assigned to the first relevance network. The algorithm then recursively descends to each of the connected features, and assigns them the same network number. This continues until all the features have either been assigned to a network, or have been eliminated.

```
define a new procedure called assign-network, that takes a
      current-gene-index and a current-network-number
set G to be the gene indexed by current-gene-index in the
      array all-genes
if G's entry in the array node-connections is greater than zero
      (i.e., that node is being used in a connection)
if G has not been assigned a network number
      assign it the current-network-number
else
      just in case we find a gene that was already
            assigned a number,
      if G's current network number is larger
            than current-network-number change G's
                  network number to current-network-number
      end if
end if
if we have assigned or changed G's network number, we need to
      do the same to G's connected genes
loop through all of G's active connections, using the
      measure-triangle, using index H
      if gene H's network number does not equal G's network number
            call assign-network recursively on gene H
                  using current-network-number
      end if
```

```
end if
end procedure

set the current-network-number to 1
to start the assignment, loop through all the genes, using index G
    if G's entry in the array node-connections is greater than zero
        (i.e., that node is being used in a connection) and G has
        not been assigned a network number
    call assign-network on G using current-network-number
    increment current-network-number
end loop
```

Finally, the results need to be displayed graphically. There are several standard formats for graph descriptions, and the specifics of these are beyond the scope of this book. We have successfully used both the GMF format used by Tom Sawyer Software, Inc. (Oakland, CA) in their Graph Editor Toolkit, as well as the DOT format used by AT&T Research Labs in their Graphviz package. For the purposes of this chapter, we will demonstrate how one of these networks may look in DOT format, primarily because software to format the graphs is freely available for academic use along with the source code.[7] One can output this type of file by first looping through all the nodes and genes in use and printing out their characteristics, then looping through and printing all the connections. A sample output file in this format is shown here.

```
graph RN {
    size="7.5,10";
    ratio="auto";
    node [ fontname = "Helvetica", fontsize = 8 ];
    node001 [ shape=box, height=0.2, width=1.6, label="1318_at",
        style="setlinewidth(1.5)"];
    node002 [ shape=box, height=0.2, width=0.8, label="36262_at",
        style="setlinewidth(0.1)"];
    node003 [ shape=box, height=0.2, width=0.8, label="422_s_at",
        style="setlinewidth(3.0)"];
    node001 -- node002 [weight=1,style="setlinewidth(0.1)"];
    node001 -- node003 [weight=1,style="setlinewidth(3)"];
}
```

---

[7] http://www.research.att.com/sw/tools/graphviz/.

After installing the Graphviz package, this sample graph can be stored in a file called "test.dot." Executing the following commands on a standard Linux computer will first translate the test.dot into a PostScript file called test.ps, which can then be processed to make an Adobe Acrobat PDF file called test.pdf.

```
dot -Tps -o test.ps test.dot
ps2pdf test.ps
```

The output PDF file from this example will contain three nodes connected with two lines. Alternatively, a commercial software package can be used to lay out more sophisticated graphs, such as the Graph Layout Toolkit. Using this package, we have constructed relevance networks such as those in figure 4.10.

Typically, several graphical conventions are used to highlight the findings in relevance networks, as shown in the figure 4.11. Since each line represents a strong connection or putative association, the thickness of each line is proportional to the strength of the dissimilarity measure, so thicker connections are stronger (and thus easier to find on the diagram). The width of each box representing a feature is proportional to the number of features to which it is connected, and the edge thickness of each box is proportional to the average strength of all the connections entering that box. The length of the connection lines is not significant, and is left variable so that the graph layout algorithms can more easily place the networks in a visually appealing manner (*e.g.*, minimizing edge crossings).

A question immediately arises: Where does one set the threshold? One way to approach this question is to plot the distribution of dissimilarity measure scores, then recompute the scores using repeatedly randomly permuted data sets (see section 4.12.1). For instance, if one randomly permutes the data set 100 times and finds the highest dissimilarity measure score is x, then one could set the threshold to be greater than or equal to x. This approach is shown in figure 4.15. However, an important point to raise is that the threshold does not have to be a fixed measurement, but instead can be "dialed" up and down. In other words, one can set the threshold to be arbitrarily high to start with and the resulting relevance networks can be viewed. If the associations are strong *but are already known and no longer worthy of pursuit*, the threshold can be "dialed down" slightly until new genes and connections are added, creating potentially novel hypotheses. Of course, one must be careful not to drop the threshold so low as to allow potentially noisy and inaccurate associations. The permutation testing can help with this.

Published examples

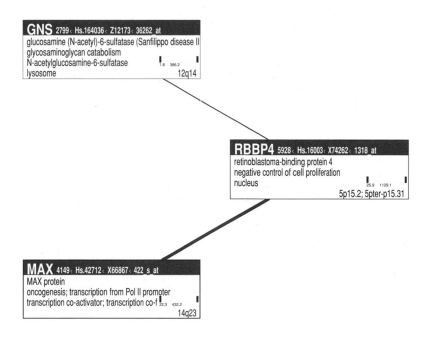

Figure 4.10
Sample relevance network. One of 78 networks formed by aggregating mouse samples from six different experiments. Threshold $\hat{r}^2$ was set to .99, meaning that the "thinnest" line seen represented an $\hat{r}^2$ of .99. The legend for this type of diagram is shown in figure 4.11. The hypothesis generated here is that *RBBP4* expression correlates with *GNS* and *MAX* gene expression. The miniature histograms for each gene show no "spiking" in the expression measurements (see section 4.5.2). Each of these genes appears at a different chromosomal location. By including information from LocusLink (see section 5.5.1) and Gene Ontology (see section 5.1.1), we see that *RBBP4* is known to control cell proliferation, and *MAX* is involved in oncogenesis.

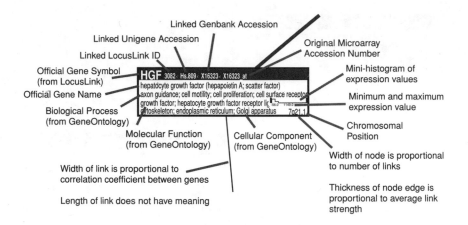

Figure 4.11
Graphical convention for relevance networks. To optimize the hypothesis generation process, relevance network display software contains ties from microarray accession number to GenBank, UniGene, LocusLink, Gene Ontology, and other databases (see section 5.5 for details of this process).

- Butte and Kohane described the methodology and first applied it to clinical laboratory data (*i.e.*, phenotypic measurements) in [37], then applied it to yeast expression measurements using mutual information as a dissimilarity measure in [38].

- Butte *et al.* [39] linked expression measurements in cancer cell lines to phenotypic measurements on those cell lines (*i.e.*, susceptibility to pharmaceutical treatment).

- Reis *et al.* [150] showed how dynamics (*i.e.*, the slope or change in expression levels over a time series) could be used as a dissimilarity measure in creating relevance networks.

- Klus *et al.* [110] constructed relevance networks from a breast cancer expression data set and linked the results to chromosomal position.

Advantages

- Captures negative correlations: Relevance networks can simultaneously display both positive and negative correlations. Negative correlations are seen commonly in biology, where greater expression of one gene (*e.g.*, a tumor suppressor gene) is associated with decreased expression of other genes.

- Genes can be connected in more than one way (genes can be connected to more than one other gene). This gives rise to three advantages: First, in biology, certain genes (*e.g.*, transcription factors) are responsible for the expression of several other genes. This type of multiple connection can be modeled using relevance networks. Second, certain features may display weak similarity to two different groups of features, and using relevance networks, these features can be seen to bridge the two groups. For example, certain anticancer drugs can perform as both topoisomerase I and topoisomerase II inhibitors, and the bridging action of these drugs between the two classes can be modeled using relevance networks.

- Multiple structures created: Certainly in biology, some gene regulatory networks are larger or smaller than others, and this variation in size can be modeled using this technique.

- Multiple measures allowable simultaneously: In theory, a relevance network could be drawn with the highest associations from multiple dissimilarity measures. For example, two genes could be linked by $\hat{r}^2$, while another two genes could be linked by mutual information.

- Multiple data types allowable simultaneously: As shown in [39], gene expression measurements can be combined with phenotypic measurements in the same data set and the resultant hypotheses easily viewed. For example, a "systolic blood pressure" measurement may appear linked to the specific genes whose expression measurement correlate with that measurement.

- Parallelizability: The most time-consuming aspect of the construction of relevance networks is the initial comprehensive precomputation of the dissimilarity measure across the set of features. This computation can be parallelized.

Disadvantages

- Excessive complexity at lower thresholds: Although it is tempting to drop the thresholds to reach a potentially novel hypothesis, empirically the number of links between the nodes already present in the graph quickly grows to an unmanageable number. Depending on the graph layout software used, these typically end up looking like large circular networks with hundreds of connections between them, such as figure 4.12. In these cases, it is advantageous that dendrograms have each gene attached to the overall structure by only a single link. One can display relevance networks in a similar manner by taking each network separately and calculating the *minimum spanning tree* by iterating through all the links from weakest to strongest and deleting those links that do not cause the removal of a node in the network. Fundamentally, however, relevance networks are most useful for identifying the subgroup interactions that are linked the strongest.

- Obscures large-scale partitional structure. Because all links that are subthreshold are discarded, some large-scale functional groupings will not be detected.

- Larger cliques cannot be found efficiently: Those portions of the relevance networks that are completely connected (*i.e.*, all the nodes in the graph subset are fully connected to each other) are known as *cliques* in graph theory, and are significant in that they represent the most "believable" part of the relevance networks. In effect, if gene A and gene B are in a clique together, it means that not only is there a strong direct association between gene A and gene B but there are also indirect associations, where gene A and gene B both are associated with gene C. These cliques are useful structures, but finding them in an automated manner is an "NP-hard" problem (*i.e.*, the size of the graph quickly overwhelms the ability of any processor to find them using brute force). [8]

- Entire vector is used: Like other analytic methods, this means that for each gene, the entire set of samples are used in calculating the dissimilarity measurements. However, there may be very little variation in most of the samples, and great variation in only a few of the samples. Two genes might be scored as being similar, when they may actually differ greatly in one of the many samples.

- This is not a globally optimal network: Early bad decisions cannot be corrected later as the threshold drops. Like dendrograms, relevance networks fall under the category of "greedy algorithms" in computer science, as the strongest links are always used at every step, regardless of the later consequences this may have for forming the networks.

- Computational requirements: This technique requires the comprehensive precomputation of the dissimilarity measure for all pairs of genes, which grows on the order of n^2, where n is the number of genes being studied. This can be a time-consuming computation. However, compared to dendrogram formation, after the precomputation, no further dissimilarity measures need to be computed, and thus with proper programming, the network formation can be performed without much memory.

- Ternary and higher comparisons: As described here, this technique performs pairwise comparisons. However, there are several examples in biology where a combina-

---

[8]See sections 1.4.1 and 5.2 for more thoughts on "NP-hard" problems.

tion of events is required for downstream action to take place. For example, several transcription factors may need to combine to form a larger complex before transcription is initiated. Performing pairwise comparisons is an expensive operation; performing three-way, four-way and higher comparisons comprehensively quickly becomes intractable.

- Display requirements: Optimally displaying relevance networks requires a large two-dimensional area (compared to a smaller columnar area for dendrograms) and requires the use of sophisticated graph layout algorithms for optimal node placement. (See figure 4.12 for an example of a particularly difficult network to visualize.)

4.10   Other Methods

Many other methods have also been used in functional genomics analysis. Only a short list is included here, with references to published biomedical works incorporating each technique, and using the hierarchy proposed at the start of this chapter:

1. Unsupervised.
   (a) Feature determination.
       i. Principle component analysis and singular value decomposition [189, 95, 68, 149, 5].
   (b) Cluster determination.
       i. Nearest neighbor clustering.
       ii. Agglomerative clustering.
           A. Fitch dendrogram algorithm [189].
           B. Self-organizing tree algorithm [92].
       iii. Divisive or partitional clustering.
           A. Matrix incision tree [108].
           B. Two-way clustering binary tree [4].
           C. Coupled two-way clustering [76].
           D. Cluster affinity search technique: [20].
           E. Gene shaving [87].
   (c) Network determination.
       i. Differential equations [44].
       ii. Bayesian networks [71].

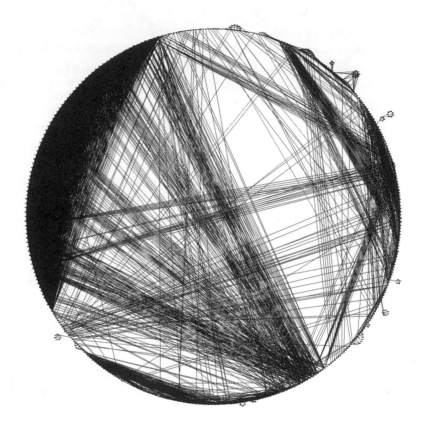

Figure 4.12
Relevance network laid out as a circle. Relevance networks with too many genes may be
rendered in circles that minimize overlapping nodes and edge crossings. However, these may be
impossible to visualize.

     iii. Hybrid petri networks [128].

     iv. Dynamic bayesian networks [134].

     v. Boolean regulatory networks [119, 197, 172, 1, 2].

2. Supervised.

  (a) Single feature or sample determination.

     i. Naive bayes classifier [21, 46].

     ii. Naive Bayes global relevance [133].

  (b) Multiple-feature determination.

     i. Decision trees: Supervised and multiple-feature determination [59].

     ii. Support vector machines [72, 33, 46].

     iii. Tree harvesting [86].

  (c) Boosting [19].

4.11 Which Technique Should I Use?

Each of the techniques described above has its own set of advantages, disadvantages, and proper uses. It is useful at this point to revisit our initial list of potential hypotheses and determine which bioinformatics technique would be most helpful to apply in each case.

- *What uncategorized genes have an expression pattern similar to these genes that are well characterized?* One could iterate through the group of well-characterized genes and apply the supervised nearest neighbor technique to find genes with similar patterns. Similarity in this context could be defined using any of the established dissimilarity measures (starting with the correlation coefficient and Euclidean distance).

- *How different is the pattern of expression of gene X from other genes?* Starting with all possible gene-gene pairs, one can first comprehensively compute a dissimilarity measure across all the pairs (*e.g.*, measuring the Euclidean distance between all pairs of genes). This set of measures will form a distribution, and very possibly a bell-shaped distribution. One can then find where the dissimilarity measures for gene X are in this distribution. If, for instance, the dissimilarity measures for gene X are in the top 5th percentile of the overall distribution, it would suggest that gene X is more dissimilar than other genes are.

- *What genes closely share a pattern of expression with gene X?* One can use a supervised nearest neighbor approach to find those genes with expression patterns most similar to gene X.

- *What category of function might gene X belong to?* Here, one can create a dendrogram, then find gene X in the dendrogram. As one progresses up the tree starting from gene X, one finds those genes with expression patterns similar to gene X. Using a "guilt by association" approach, if most of the neighbors of gene X belong to a particular functional category, then one should hypothesize that gene X belongs to that category [63]. Obviously, this approach is only useful when the functional categories are known for the majority of the genes. An alternative, simpler approach than constructing a dendrogram is to compute the nearest neighbors to gene X and ascertain what their functions are.

- *What are all the pairs of genes that closely share patterns of expression?* Although constructing a dendrogram starts with the comprehensive pairwise computation of a dissimilarity measure, these pairs are not clearly displayed in the results. Instead, one can use relevance networks to find those pairs of genes that score highest with a dissimilarity measure.

- *Are there subtypes of disease X discernible by tissue gene expression?* To answer this question, one can easily construct a dendrogram across the samples, not across the genes. For instance, depending on the mixture of samples collected, one might find that half of the samples fall in one half of the dendrogram, and the other half of the samples fall in the other half of the dendrogram. One can then proceed to determine if there are phenotypic differences between these branches. This was most dramatically demonstrated by Alizadeh *et al.* [3], where the two major branches in the gene expression dendrogram obtained were found to correspond to patients with significantly different mortality (see figure 4.13). Of course, if the sample mixture is not diverse enough, one might miss the true biological subsets and might instead overfit normal variance or noise in the measurements. This is a matter of experimental design addressed in chapter 2.

- *What tissue is this sample tissue closest to?* One can create a dendrogram of the tissues and determine where the unknown tissue lies. Alternatively, one can use a supervised nearest neighbor approach to determine the "closest" pattern of gene expression.

- *What are all the different patterns of gene expression seen?* Although dendrograms can provide a categorization and ordering of genes based on expression pat-

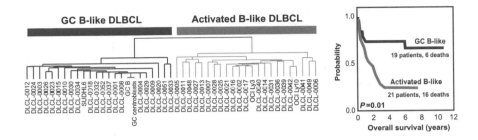

Figure 4.13
Subcategories of B-cell lymphoma determined by microarrays correspond clinically to duration
of survival. On the left is a dendrogram constructed across the samples of B-cell lymphoma,
using an unsupervised technique. The top branch essentially defines an even split between the
categories GC B-like DLBCL and Activated B-like DLBCL, but this distinction was never before
made clinically. On the right are Kaplan-Meier survival curves of the patients from whom the
samples were obtained. Patients whose cancer matched the Activated B-like DLBCL gene
expression profile had a significantly worse prognosis. (From Alizadeh *et al.* [3].)

tern, self-organizing maps can easily categorize and summarize the expression patterns present. Further analysis can be run on these clusters, including determining
whether the genes fall in the same functional category, or searching for common
sequences in the 5′ upstream region from each gene.

- *Which genes have a pattern that may have been a result of the influence of gene X?*
  Answering this question requires the proper biological experiment to be performed.
  Thus, this question is often asked when gene X is being over- or under-expressed
  in a particular sample. One can treat the controlled expression level of gene X as
  an environmental factor, then, using a supervised nearest neighbor approach, one
  can find whether any of the genes correlate with the controlled expression level.
  However, the controlled expression level may not be quantifiable, but instead may
  be discrete (*e.g.*, gene X is "absent" in a sample from a knockout animal, but
  is "present" in a sample from a wild-type or normal animal). In this case, one
  can create a hypothetical gene expression matching this information (*e.g.*, high
  expression levels in the one set of samples, low expression levels in the other set),
  then use a supervised nearest neighbor approach to find genes with a similar pattern.

- *What are all the gene-gene interactions present among these tissue samples?* Relevance networks provide a network of gene-gene interactions based on dissimilarity
  measures.

- *Which genes best differentiate two known groups of tissues?* Because the two groups
  of tissues are known, this calls for a supervised approach. Classic methods can be

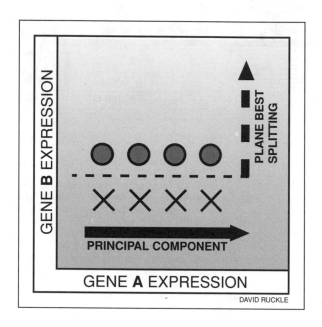

Figure 4.14
Feature reduction with principal components analysis. Two genes are measured in eight samples, four from one disease (marked by X) and four from another disease (marked by O). The expression measurements of gene A are represented on the x-axis and the expression measurement of gene B are represented on the y-axis. Note that the first principal component (*i.e.*, the vector that captures the most variance) is parallel to the x-axis, which corresponds to gene A. However, the line that best splits the two diseases is described by its orthogonal vector along the y-axis, which corresponds to gene B. This means that although the variance seen in gene A best explains the overall variance seen in gene expression measurements in each disease and both together, the variance seen in gene B is best used to split the two diseases. Although this is easy to distinguish in this simplified two-dimensional case, it is not so easily visualized in a real-world, multidimensional data set.

used to address this question, such as the Student's t-test. Alternatively, one can construct a hypothetical gene expression pattern across the samples with the desired behavior (*e.g.*, high expression levels in one tissue, low expression levels in the other tissue), then use a supervised nearest neighbor approach to find genes with a similar pattern. [78] Newer approaches to answer this question involve the use of support vector machines [33, 49, 72]. It is important to note that simplifying or pruning the list of genes using a technique such as principal components analysis does not give one the genes that best split two sets. A classic example of this is in shown in figure 4.14.

- *Which gene-gene associations best differentiate these two groups of tissue samples?* Answering this question requires a data set with sufficient measurements from the two types of tissues, and the analysis involves a combination of approaches. First, for each of the two tissue samples, one can first comprehensively compute a dissimilarity measure for all the gene-gene pairs. Then one can use a supervised approach to compare the dissimilarity measure for the two tissue types. [9] For example, if the expression pattern correlation coefficient between gene A and gene B was .95 in breast cancer, but was .25 in lung cancer, one might make a biological hypothesis that there is a difference in this interaction between the two diseases.

- *Which genes have a pattern similar to this organismal, phenotypic, or environmental factor?* If one can treat the external factor as a hypothetical gene expression pattern, then one can apply the supervised nearest neighbor technique to determine genes most similar to that pattern.

4.12   Determining the Significance of Findings

One traditional way to gauge whether the findings for a particular microarray analysis are significant is to establish whether statistical significance is present. Traditionally, this involves the use of particular algorithms or formulas to determine the likelihood that that particular finding was governed simply by random chance. For example, in traditional clinical trials, when a study finding is noted to have a statistical significance, or p value of $< .05$, the assumption is that the likelihood of the findings in that particular analysis being due to random chance is less than or equal to 5 chances in 100.

These analyses are problematic when applied to functional genomic data sets. Suppose we have 10 tissue samples from patients with disease and 10 samples from patients without the disease. For a single gene, we can estimate how likely it is that the observed difference in the means of the expression of that particular gene in the two sets of samples could be attributable to chance. A simple lookup in the probability density function for that statistical test (*e.g.*, a t-test) will provide this estimate. However, if the p value is found to be $< .05$ for that particular gene, then in a gene expression microarray measuring the expression of 10,000 genes, this would mean that we would expect to find the observed difference between the

---

[9]That is, we compare the measure-triangles for each tissue. The elements of the measure-triangle that are most different will suggest different relationships between the gene pairs corresponding to each of these dissimilar elements of the measure-triangle. Because the whole measure-triangle is used, the result is comprehensive.

mean expression values across the two sets of 10 samples for 500 genes solely due to chance.[10] Determining how to obtain the significance threshold for all the different bioinformatics techniques we have described above is an active area of research but has yet to make its way into the mainstream of published microarray analyses. Instead, pragmatic tests of the robustness of the results found have been adopted of the form described below.

4.12.1   Permutation testing

With the large number of comparisons being made, a valid criticism should be raised that perhaps high correlation coefficients or extreme values in other dissimilarity measures are found due to chance. One way to counter these arguments is to estimate how often random chance would generate correlation coefficients of the magnitude obtained in the original analysis. Typically, this is done through use of permutation testing.

One can take the entire array of cases and features (*i.e.*, samples and genes, respectively) and randomly shuffle the values in each column of features, thus breaking the links between the features. The mean and standard deviation expression level for each gene would remain the same. One can then repeat the entire analysis (*e.g.*, dendrogram or relevance network) that one is performing 10 or 100 times or more to determine the strongest dissimilarity measure calculated.

For example, let us assume we are analyzing a microarray data set with 60 cancers and 2000 genes measured in each of those cancers. We want to perform a dendrogram analysis on this data set. As before, we would start by calculating the comprehensive pairwise comparisons. Within this large set of comparisons, are high correlation coefficients likely to be spuriously generated? One way to answer this is to take each of the genes and randomly shuffle the gene expression measurements among the cancers and then repeat the entire comprehensive pairwise calculations. This permutation and calculation of the measure-triangle is repeated several times. For each iteration, the distribution of $\hat{r}^2$ values is stored. After all the iterations we will have a distribution of correlation coefficients from the original data set as well as several or an averaged distribution of correlation coefficients in the permuted and randomly shuffled sets. This is shown in figure 4.15.

If there is structure in the expression data set that is more robust than what would be expected by chance, the distribution of correlation coefficients from the original analysis should be wider than the distributions of the randomly permuted

---

[10]Many assumptions go into this estimate, including clearly erroneous ones such as the independence of all genes in their expression patterns.

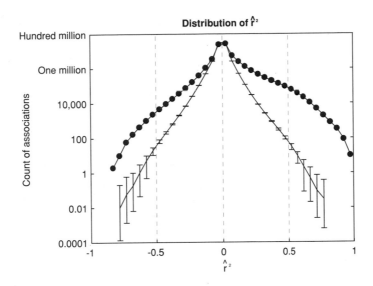

Figure 4.15
Distribution of $\hat{r}^2$ in the original versus permuted gene expression data set. The distribution of $\hat{r}^2$ calculated using an original gene expression data set is shown with solid circles. For each gene, expression measurements were independently randomly shuffled 100 times. The average distribution of $\hat{r}^2$ is shown with error bars covering 2 SD. In this example, random permutation was unable to create an association with $\hat{r}^2 > 0.80$ or < -0.85. (From Butte *et al.* [39].)

sets. We view this difference in area-under-the-curve between the two distributions as a measure of the signal-to-noise in the analysis. If the distribution of correlation coefficients in the randomly permuted data sets is as wide as the distribution correlation coefficients from the original unshuffled data set, it is a clear indication that the high correlation coefficients being seen could very well be due to random chance.

This permutation testing does not substitute for a p value, as traditionally used in tests for statistical significance, but it is one generally accepted way (in the functional genomics literature) to test a novel clustering algorithm and its application to determine whether findings are otherwise significant.

4.12.2   Testing and training sets

Perhaps the best way to test a model, whether constructed in a supervised or unsupervised manner, is to evaluate that model on a new set of inputs. This is typically done by determining *a priori* those inputs that will make up the training set (*i.e.*, the input samples used to create the model), and those that make up the testing set (*i.e.*, those that will be used to test the model). It is important to determine whether population-measurable characteristics are similar between these two sets. For instance, one may want to train a system using samples from one type of tissue, and test it using another type of tissue.

This type of testing is crucial in that it prevents the generation of models that overfit the original input data. It is easy to describe overfitting with an example. Imagine a thought experiment where one takes a list of tyrosine kinases (*e.g.*, insulin receptor, IGF-1 receptor) and phosphatases (*e.g.*, PP1, PTP1B). Each one is written on a flashcard, along with one of the two category names. Now, one shows the flashcards to a child repeatedly. When later quizzed with cards without the category names, the child may be able to properly assign the function just by noticing that within the input set, whenever the name contained the word "receptor," the card was referring to a tyrosine kinase. Needless to say, the "rule" this child generated is quite specific to the particular set of input samples and their representation on flashcards.

Because so many features are typically present in microarray data sets, and because there are so few categories into which the samples have to be classified, it becomes very easy for a machine-learning mechanism to generate large numbers of "rules" specifying how each feature can accurately select the proper category. However, the characteristic of these genes may be specific only to the input sample set at hand, and those rules may not be generalizable to other input samples. Thus, it becomes important to test these rules using independent inputs.

At the risk of redundancy, we emphasize that finding test and training sets that are highly comparable is of paramount importance. In the discussion of the sources of noise in microarray experiments (section 3.2), we demonstrated that we could cluster a set of leukemia experiments into their original test and training sets because of systematic differences in the measurements in the test and training sets (figure 4.16). In the presence of such systematic biases, the inaccuracy or accuracy of a clustering or classification technique on a test set may not be a meaningful measure of its performance.

Cross-validation testing   Cross-validation is a classic approach to testing for robustness in models generated by machine-learning algorithms, built on the idea of having both a testing and training set, but for use when the total number of inputs is small. This technique works by repeatedly partitioning the available input into subsets for repeated trials. For each of these subsets, the algorithm or machine-learning technique is run and the output model is saved. After all the trials, the set of output models can be analyzed to determine those characteristics or features

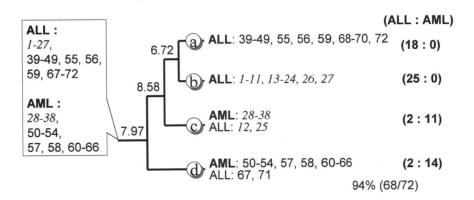

Figure 4.16
Identifying differences between the test and training sets through clustering. The matrix incision tree algorithm from [108] was applied to the leukemia classification problem published by Golub *et al.* [78]. The algorithm correctly clustered 64 of the overall 72 cell lines (94%) in the data set, placing the acute myelogenous leukemia (AML) samples in branches (c) and (d), the acute lymphocytic leukemia (ALL) samples in branches (a) and (b), and misclassifying four ALL samples into branches (c) and (d). However, the matrix incision tree also successfully revealed the distinction between the published training set (cases 1–38, italic, placed in branches (b) and (c)) and the test set (cases 39–72, placed in branches branches (a) and (d)) with 100% accuracy. It is unlikely that the biology of the test set of the AML and ALL cell lines were that different from the cell lines in the training set. It is much more likely that systematic changes in the hybridization conditions, or different sample preparation, or the use of a different batch of microarrays were responsible for the clustering of test versus training sets.

that are consistently represented across the trials. As an example, if in 10 trials, gene A is always found to accurately classify a set of samples between two diseases, but gene B accurately classifies in only 6 of the 10 trials, then gene A may be viewed as a more robust classifier, in that it may be less influenced by particularities of a single input set.

The traditional way this is performed is using *n-way leave-one-out cross-validation*, where n is the number of input samples: n input subsets are created, each missing the nth sample, then the machine-learning technique is applied to each of these input subsets. Cross-validation testing is easier to apply to supervised learning methods than to unsupervised, as it is difficult to determine accurate metrics for deciding on the strength of the generated unsupervised models. This approach has been used in several published works, including [19, 33, 46, 60, 72, 78, 86, 90].

4.12.3 Performance metrics

In evaluating genes or features that are meant to decide one category versus another, other relevant performance metrics include sensitivity, specificity, positive predictive value (PPV), accuracy, and area under the receiver operating characteristic (ROC) curve [79].

Let us refer to the state of disease or any category of interest as the "disease state" and all other categories or states as "nondisease states." The sensitivity of a test measures the proportion of correctly categorized disease cases out of the total number of actual disease cases, while specificity measures the proportion of correctly categorized nondisease cases out of the total number of actual nondisease cases. Positive predictive value is the fraction of cases correctly categorized as disease over the number of all disease cases, regardless of categorization. Accuracy is calculated by the ratio of correctly categorized cases over all the cases.

To further illustrate these metrics, figure 4.17 shows a contingency table in which the two classes are called 0 (non-disease) and 1 (disease). The vertical groupings of cases are gold standard-labeled[11] cases belonging to each class. The horizontal groupings may be cases automatically categorized into each class. For example, a cases and c cases were actual class 0 cases, though a cases and b cases were labeled by a particular model as class 0. Similarly, b cases and d cases were actual class 1 cases, while c cases and d cases were labeled by a particular model as class 1. The relevant performance metrics in equation form are then as follows:

---

[11] For example, defined classes of two different types of leukemia as judged by an expert panel.

Classified as:	Actual class: 0	Actual class: 1
0	a	b
1	c	d

Figure 4.17
Contingency table illustrating performance metrics for classification algorithms.

$$\begin{aligned}
\text{Sensitivity} &= \frac{d}{b+d} \\
&= \frac{\text{Number of correctly categorized event cases}}{\text{Total number of event cases}} \\
\text{Specificity} &= \frac{a}{a+c} \\
&= \frac{\text{Number of correctly categorized nonevent cases}}{\text{Total number of nonevent cases}} \\
\text{PPV} &= \frac{d}{c+d} \\
&= \frac{\text{Number of correctly categorized event cases}}{\text{Number of categorized event cases (correct and incorrect)}} \\
\text{Accuracy} &= \frac{a+d}{a+b+c+d} \\
&= \frac{\text{Number of correctly categorized and nonevent cases}}{\text{Total number of cases}}
\end{aligned}$$

4.12.4 Receiver operating characteristic curves

The ROC curve is a plot of sensitivity versus one minus specificity. Because sensitivity and specificity can be inversely varied by altering the threshold at which a case is categorized as one class or the other, the *area* under the ROC curve more effectively describes a classification algorithm's discriminatory ability. Stated differently, for a particular implementation of a classification algorithm, the trade-off between sen-

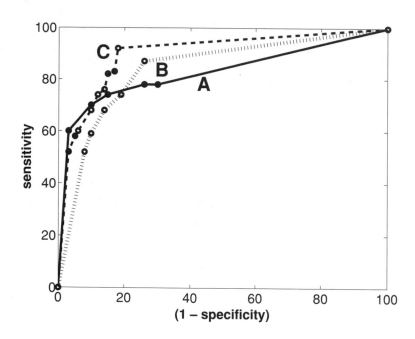

Figure 4.18
Sensitivity and specificity trade-offs are defined by the ROC curve. ROC curves A, B and C describe the trade-offs in sensitivity and specificity for the same data set using three different classification algorithms. C is clearly the best algorithm because the area under the curve is the highest. That is, it provides a better set of sensitivities and specificities for all values of these. In contrast, B and C provide trade-offs inferior to A and for some range of sensitivities, B provides better performance than C, and *vice versa*.

sitivity and specificity is fixed as specified by the ROC curve. To find a different trade-off, one must obtain a different classification algorithm implementation that produces a different ROC curve. This is elaborated upon in figure 4.18.

To calculate the ROC curve for a single continuous variable[12] all that is required is to obtain the sensitivity and specificity at a sufficient number of thresholds. The procedure is somewhat more complex with classification models that involve different thresholds on several variables. For example, for decision trees, ROC curves are determined by first assigning to each tree leaf the probability of being an event for a set of derived values that percolates to that point. These probabilities are based upon the ratio of events to nonevents that fall into each leaf during

---

[12]For example, if gene A > threshold, then the prediction is that the sample is from a patient with the disease, and if gene A ≤ threshold, then the prediction is that the sample is from a patient without a disease.

training. The threshold for considering a case to be event or nonevent is then set at each leaf probability value. The resulting sensitivity-specificity pairs, when plotted on a grid of sensitivity versus (1-specificity), gives the ROC curve.[13]

4.13   Genetic Networks

It is a testament to the ambitions of bioinformaticians that among the principal early goals they had for the use of microarray data was nothing less than to *reverse-engineer* the genetic networks responsible for maintaining cellular function. Although these ambitions may have been somewhat premature, we include them here because the goal of elucidating genetic networks remains a major challenge and problem of interest in this field.

4.13.1   What is a genetic network?

As covered in section 1.5, every human cell has on the order of 10^4 genes. Typically, not all the genes are expressed equally at any one time. At different stages of development, or as a response to extracellular perturbation, different genes of a cell are switched on at different intensity levels and at various time points. The genes and their associated, generally time dependent, levels interact with one another in a sophisticated regulatory web of cause and effect known as a genetic network. Restated, genetic regulatory networks are the set of mutually activating and repressing genes and gene products and their interactions [64].

It is generally difficult, if not impossible, to come up with the full set of parameters which completely characterize a genetic network. These parameters may be *intracellular* in nature, such as the mRNA concentration of another gene; *intercellular*, such as a hormone secreted by one organ to communicate a signal to another; or *extracellular*, such as an environmental condition like heat shock. Uncovering and understanding these interactions may potentially lead to new applications and insights on developmental disorders and diseases such as diabetes, cancer, and others.

The expression level of a gene is manifest by the presence of mRNA corresponding to its DNA transcript. Qualitatively, it is commonplace to say that a particular gene is highly expressed when high levels of its complementary mRNA are detected, whether by Northern blot, PCR, or microarray. Quantitatively, the reader may ask

---

[13]However, because the number of leaves in a decision tree are finite, the number of points that can be ascertained for that ROC curve is correspondingly finite and therefore the curve appears as a segmented trajectory of straight lines.

the obvious and more delicate question of how much mRNA should be present in a system in order that the gene be critically expressed so as to effect a biological state transition. In practice, this question is made more difficult by the realities of detection device sensitivity, repeat measurement variations, and how one defines a measurable biological state transition.

With the advent of microarrays, it has become possible to measure, in a relatively accurate and easy way, the expression levels of a great number of genes simultaneously. As a consequence, it is now possible to investigate topics such as time course development and drug intervention within the framework of the coordinated action of an entire transcriptome—the cellular genetic network—with the objective of reverse-engineering or unraveling the system's internal interdependencies.

4.13.2   Reverse-engineering and modeling a genetic network using limited data

Presently, microarray technology cannot yet measure the expression levels of all genes of an arbitrary biological system. Furthermore, there is the measurement technology-independent complication noted earlier such that it is generally impractical to obtain the full set of parameters that completely determine the state transitions of the system. Here, in essence, the reverse-engineering of a typical genetic network would mean uncovering dependencies and causal relationships between genes using a limited data set or knowledge base. Data in this context include empirical measurements of the respective components, in particular the mRNA levels, as well as *a priori* biological facts about the system under investigation. The information is limited in the sense that we do not have corresponding data from every gene present in the system and, furthermore, the data we do have inevitably contain measurement or experimental errors. An example: Suppose that a researcher wants to study the interaction between two mouse genes, gene A_1 and gene A_2, in liver from the day of birth until 1 week later. The researcher will use a robotically spotted microarray with probes for genes A_1 and A_2. As samples, she or he harvests liver cells at N uniformly spaced time points from postnatal day 1 to day 7. Let us consider the possibilities of interaction or regulation between A_1 and A_2, as discussed in [164]:

1. A_1 regulates A_2, or *vice versa*. This interaction may be indirect, propagating through an intermediate cascade of other genes which are not being assayed.

2. A_1 and A_2 are co-regulated by another gene or a subset of genes.

3. A_1 and A_2 do not interact with one another nor do they share a common co-regulator.

The term *regulate* here carries causal implications. One way to translate the biological statements above into mathematical formalism is to say that (1) and (2) are equivalent to saying that A_1 and A_2 are not stochastically independent, whereas (3) says that A_1 and A_2 are stochastically independent. An immediate question that comes to mind is whether it is technically possible to infer causal relationships between genes by their expression level data alone. For such a scenario, the most typical mathematical tool that our researcher could apply is the linear or rank correlation coefficient, r—provided that the number of time point measurements at which a gene is assayed, N, is large enough,[14] for if $N = 2$ then any two genes are perfectly (and vacuously) correlated. As noted in [164], the observation of correlation between pairs of variables is commonly used in biology to predict causal relationships which have to be validated biologically. Indeed, that observation drives much of the data-mining machinery discussed in this chapter.

From a purely mathematical standpoint, correlation cannot be used to rigorously prove or disprove the existence of such relationships. Recall from our earlier review section on correlation (section 3.6.1) the basic rule of thumb that while independence implies zero correlation, the converse is not generally true. In spite of this, a high enough correlation (say $|r| > 0.9$, by no means a formal threshold) between genes A_1 and A_2 is a reasonably strong argument for further laboratory investigation of the pair, after taking into account the size of N, measurement errors in the primary data, and established facts about the pair's biological background.

Another approach to analyzing a genetic network is to view the whole as a dynamical system, not just consisting of pairwise interactions, and to take advantage of the traditional systems tools that already exist, (see [80] and [58]). Suppose that our researcher measures the expression level for genes G_j $(j = 1, \cdots, M)$ at times t_i $(i = 1, \cdots, N)$, denoted by $x_j(t_i)$. We also suppose that the expression level of a gene A_k at time t_q $(q > 0)$ is a linear combination of the expression levels of all other genes G_j from the previous time step t_{q-1} [or more generally from all previous time steps t_r $(0 < t_r < t_q)$],

$$x_k(t_q) = \sum_{j=1}^{M} w_{k,j} \, x_j(t_{q-1}) + \beta_k \qquad (4.13.1)$$

---

[14]In practice, sufficiently high-resolution time series of microarray measurements performed over an adequate interval are rare if not absent.

or rewriting in a matrix format,

$$
\begin{pmatrix} x_1(t_q) \\ x_2(t_q) \\ \vdots \\ x_M(t_q) \end{pmatrix} = \begin{pmatrix} w_{1,1} & w_{1,2} & \cdots & w_{1,M} \\ w_{2,1} & w_{2,2} & \cdots & w_{2,M} \\ \vdots & \vdots & \cdots & \vdots \\ w_{M,1} & w_{M,2} & \cdots & w_{M,M} \end{pmatrix} \begin{pmatrix} x_1(t_{q-1}) \\ x_2(t_{q-1}) \\ \vdots \\ x_M(t_{q-1}) \end{pmatrix} + \begin{pmatrix} \beta_1 \\ \beta_2 \\ \vdots \\ \beta_M \end{pmatrix}
$$

where $w_{k,j}$ is a weight term encoding the influence of the expression of gene G_j on the expression of G_k, and β_k is a parameter which may be stochastically distributed which could encapsulate *a priori* knowledge about regulation. Note that EQ. (4.13.1) is linear in the expression quantities and is first-order Markov, *i.e.*, the state of the system at time t_q is entirely determined by the time t_{q-1} state. We could also write EQ. (4.13.1) as a difference equation in keeping with our systems viewpoint,

$$
\Delta x_k(t_{q-1}) = \sum_{j=1}^{M} w'_{k,j}\, x_j(t_{q-1}) + \beta_k
$$

$$
w'_{k,j} = \begin{cases} w_{k,j} - 1, & \text{if } k = j, \\ w_{k,j}, & \text{otherwise,} \end{cases}
$$

(4.13.2)

where $\Delta x_k(t_{q-1}) \doteq x_k(t_q) - x_k(t_{q-1})$. Unlike the classic system of difference equations, $w_{i,j}$ here are the unknowns and $x_j(t_q)$ are the known measured quantities. In general, $w_{k,j} \neq w_{j,k}$. Recall from linear algebra that in order to solve for the M^2 unknowns $w_{i,j}$ uniquely, one needs to have M^2 different equations EQ. (4.13.2); in other words, there should be enough data time points. If the number of unknowns exceeds the number of equations, then the system is underdetermined and may have infinitely many solutions for $w_{i,j}$. Conversely, if there are more unknowns than there are equations, then the system is overdetermined and it may be algebraically insoluble. In this situation, one can attempt to find a weight matrix $w_{i,j}$ that "solves" the system in the sense of minimizing an *ad hoc* energy functional. A typical example of this approach is the least squares fitting of data to a line where the energy functional is the sum of distances from the fitting line. Once the weights $w_{k,j}$ in EQ. (4.13.2) have been computed and we suppose as modelling assumption that $w_{k,j}$ are time-independent, then EQ. (4.13.2) is a well-defined difference system and one may apply standard dynamical systems or Markov chain techniques to explore interpolation and prediction-type questions such as:

1. As a reality check: Does the computational result or prediction agree with the outcomes from the original physical experiment?

2. How would the model EQ. (4.13.2) evolve under a different set of initial states— simulating a different experimental setting? Are there particular initial conditions which lead to dramatically different subsequent behavior in the presence of a small perturbation? Continuous dependence on initial state or bifurcations?

3. If we removed the time-series data for a gene G_k and recalculate the weights $w_{i,j}$, how different is the resulting model from the one with G_k present?

4. What is the asymptotic or steady-state behavior of EQ. (4.13.2)? Attracting sets, hysteresis cycles, recurrent states?

The state of a system at any time t_q refers to the collective expression profiles $x_k(t_q)$ for $k = 1, \cdots, M$. In a developing hippocampi system modeled with EQ. (4.13.2), Somogyi *et al.* found that the weight matrices were sparse *i.e.*, most of the entries were 0, in agreement with the generally accepted intuition that genes are not regulated equally by every other gene [164]. From a computation-theoretic point of view, it may also be interesting to study the time-continuous analog of the difference equation (4.13.2), which would be a system of linear differential equations,

$$\frac{dx_k(t)}{dt} = \sum_{j=1}^{M} w'_{k,j}\, x_j(t) + \beta_k$$

or stochastic differential equations, if we allow β_k to be a specific probability distribution as in [109]. When one has a deterministic system of linear ordinary differential equations, Fourier or wavelet methods could be used to transform into a system of algebraic equations.

An attendant complication with these methods is the aliasing phenomenon. One could potentially be undersampling the respective RNA levels below the fundamental frequency (probably cell cycle-related) of the biological system. As is well-known in signal processing, it is not possible to reconstruct the originating signal from samples that were obtained less than twice the threshold rate known as the Nyquist frequency. The model can easily be generalized to become a nonlinear one by letting $w_{j,k}$ be both a function of time and expression, or to a higher-order Markov by assuming that the expression level of A_k at t_q depends upon the expression levels of A_j's for other previous times t_r, $0 < t_r < t_q$. Weaver *et al.* [186] and Wahde and Hertz [185] studied a modified version of EQ. (4.13.1),

$$x_k(t_q) = g\left(\sum_{j=1}^{M} w_{k,j}\, x_j(t_{q-1}) + \beta_k\right)$$

where $g(y) = \left(1 + e^{\lambda y}\right)^{-1}$ is a modified step function where $\lambda > 0$ is the rate that the value of g changes from 0 to 1 at $y = 0$. When $\lambda \uparrow \infty$, $g(y) = 0$ (resp. 1) for $y < 0$ (resp. $y > 0$).

There are several biologically based reasons to motivate this step function-like form of g (*e.g.*, upon binding of a transcriptional repressor or initiation of a kinase cascade). Note that g has a well-defined inverse function, g^{-1}. The weights of the system above can be solved via linear algebra after applying g^{-1} on both sides of the equation. Operating within the systems paradigm, Hanke and Reich [83] and Somagyi and Sneigoski [164] have also modeled Eq. (4.13.2) as neural networks which have the advantage that one could train the parameters—in this case, the weights $w_{i,j}$—to match the data set. Potential issues here include the selection of the type of neural network that is optimally suited to the questions being studied, the length of time required for training, and the characteristics of the training set.

Thus far we have considered the expression levels x_k to be essentially continuous variables interacting in a linear combinatorial manner. One may as validly let x_k be a discrete-valued variable taking values 0 and 1, where 0 indicates that gene A_k is off and 1 on, to produce a Boolean network as in [119]. State transitions are captured by a truth table which we have to discover using *ad hoc* rules optimizing mutual information.

Recently, there has been active work in the area of analyzing genetic networks within a probabilistic framework; of particular note is the application and development of bayesian models of gene expression and interaction which will be detailed in the next section.

It probably bears repeating here that no mathematical treatment of observational data will be sufficient to be convincing about genetic causal relationships. For that to occur, the tried-and-true methods of scientific hypothesis testing (see figure 2.2) and the concomitant "wet" laboratory work will remain necessary. Therefore it should not be too disappointing to note that we have still not adequately addressed the issue of how one may infer causal relationships from expression data, in particular limited expression data. From the purely computational point of view, it is not possible to do this without the aid of established biological facts and experimental intuition. This in turn requires a productive functional genomic pipeline of the sort diagrammed in figure 1.3.1.

4.13.3 Bayesian networks for functional genomics

Bayesian networks (BBNs) [139], also known as causal probabilistic networks, represent one of the most successful advances of artificial intelligence (AI). Born at the confluence of statistics and AI, BBNs provide a sound way to discover models of interaction among variables. When applied to functional genomics, they offer the opportunity to discover complex pathways of activation and interaction among genes. Although still in an early stage, the applications of BBNs to functional genomics already displays the common problems statistical methods encounter in the realm of functional genomics: the large number of variables met by a small number of cases and the consequent under-determination of the models learnable from data. Nonetheless, because BBNs provide at least the potential of determining genetic networks on the basis of probability theory, and therefore all the desiderata that flow from such a grounding, they are beginning to attract increasing interest in this field. This section first reviews the basic concepts underpinning BBNs and then describes the early attempts to apply this methodology to functional genomics.

Bayesian networks   Bayesian methods provide a sound and flexible framework to reason about the interplay between conditions and effects under uncertainty. The theoretical core of bayesian theory itself—the well-known Bayes' theorem— is built around the intuition that the path leading to an effect can be reversed and a probability can be assigned to the set of possible conditions responsible for that effect [148]. The resulting approach also provides a straightforward way to integrate human-derived domain knowledge and domain information coming from field operations (*i.e.*, empirical biology) or simulated data.

The bayesian approach shares, with all sound knowledge representation formalisms based on probability theory, two main drawbacks. The first is representational in nature and arises from the fact that a naïve encoding of knowledge as a joint probability distribution of all the variables in the domain is very difficult to acquire from human experts. The second is computational, but is still based on the same naïve interpretation of the probabilistic representation: The probability distribution over the domain variables, so difficult to acquire from experts, can grow so large to be very costly to store and very expensive to use for reasoning.[15]

Blending together probability and graph theory, BBNs provide a way to render the probabilistic representation of knowledge representationally amenable and computationally feasible. A BBN uses conditional independence assumptions encoded by a Direct Acyclic Graph (DAG) to break down an otherwise unmanageable joint probability distribution over the domain variables into a set of smaller components, easier to define and cheaper to use. In a BBN, the nodes in the graph represent variables and directed arcs represent stochastic dependencies among the variables. As a simple example, consider the scenario displayed in figure 4.19 composed of

---

[15]See the discussion of the appetite for hard-to-get probabilities in Szolovits and Parker [173].

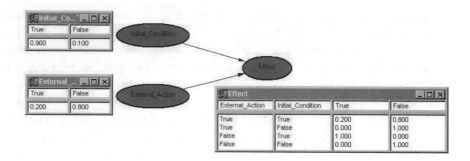

Figure 4.19
Example of a bayesian network. This network describes the impact of Initial_Condition and External_Action on Effect. This network is equivalent to the underlying joint probability distribution over the three variables, shown in table 4.1.

three variables, Initial_Condition, External_Action, and Effect, each having two states "True" and "False." The joint effect of the the two parent variables on the Effect is represented by the DAG in figure 4.19, with the directed links from the variables Initial_Condition and External_Action pointing to the variable Effect. Following the direction of the arrows, we call the variable Effect a child of Initial_Condition and External_Action, which become its *parents*.

In particular, figure 4.19 shows a BBN representing the combined action of a quite frequent Initial_Condition (90% chance of being true) and a rarely taken External_Action (20% chance of being true) over an Effect. The probability table associated with the node Effect shapes the interplay between Initial_Condition and External_Action so that, when only the Initial_Condition is in place, then Effect will be present for sure, unless External_Action is deployed to dilute the impact of the Initial_Condition. This mechanism is described by the fact that, when Initial_Condition is true and External_Action is false, the probability that the Effect is true is 1, whereas when Initial_Condition is true and External_Action is also true, then the probability that the variable Effect is true is only 0.2. On the other hand, when the Initial_Condition is not in place, then the Effect is always absent, regardless of the presence or absence of the External_Action. The BBN in figure 4.19 decomposes the joint probability distribution of the three variables in table 4.1 into three probability distributions, one for each variable in the domain. This decomposition is the key factor for providing both a verbal and a human-understandable description of the system, like the one used for the BBN in figure 4.19, and for efficiently storing and handling this distribution, which grows exponentially with the number of variables

Initial_Condition	External_Action	Effect	
True	True	True	0.036
True	True	False	0.144
True	False	True	0.720
True	False	False	0.000
False	True	True	0.000
False	True	False	0.020
False	False	True	0.000
False	False	False	0.080

Table 4.1
Joint probability distribution over the domain variables

in the domain.

   Notwithstanding these differences, both the BBN and the joint probability distribution represent the same stochastic model of the combined effect of Initial_Condition and External_Action. The BBN can therefore be used to perform all the kinds of reasoning operations provided by a probabilistic representation of the system. As mentioned before, bayesian theory provides naturally a variety of methods to reason about the behavior of a system and to understand the interplay between its conditions and its effects. Assuming sufficient microarray data (a difficult assumption at present, but likely to become less so in the near future), this is sufficient to model cellular physiology.[16] Given a set of initial conditions, a BBN can forecast their effects (prediction—particularly useful for hypothesis testing). Conversely, a BBN can walk backward a chain of dependencies and determine the probability of the initial condition given the observation of an effect (explanation—essential in understanding complex genetic networks). Even more interesting, a BBN can identify the best configuration of initial conditions to achieve a particular effect (optimization—again useful for generating testable hypotheses as well as identifying pharmacological targets), and can identify the individual contribution of an initial condition to the realization of an effect. This ability to propagate information along the network without a prespecification of the inputs allows the analyst to "drop" evidence and conjectures into the network and figure out the effect of

---

[16]With the caveats about false reductionisms articulated in section 1.4.1.

of BBNs to functional genomics is, on the other hand, very recent. BBNs hold the promise of answering very interesting questions in functional genomics and, in principle, they seem to be the right technology to take advantage of the massively parallel analysis of whole-genome data to discover how genes interact, control each other, and align themselves in pathways of activation.

BBNs offer a different view than do the more popular clustering algorithms currently used for the analysis of massively parallel gene expression data [63, 167]. While these algorithms attempt to locate groups of genes that have similar expression patterns over a set of experiments to discover genes that are co-regulated, BBNs dive into the regulatory circuitry of genetic expression to discover the web of dependencies among genes.

The promise of BBNs in functional genomics goes even further, as intensive research efforts have been addressed, during the past decade, to define conditions under which BBNs actually uncover the *causal* model underlying the data [140, 89]. The most ambitious question is therefore the following: Given a set of microarray data, can we discover a causal model of interaction among different genes? The challenge is the common problem of sound statistical methods when faced with microarray data: a large number of variables with a small number of measurements. In the context of BBNs, this situation results in the inability to discriminate among the set of possible models as the small amount of data is not sufficient to identify a single most probable model.

Friedman *et al.* [71] address these problems using partial models of BBNs and a measure of confidence in a learned model. The strategy they follow is to search a space of underspecified models, each comprising a set of BBNs, and select a class of models rather than a single one. They also adopt a measure of confidence based on bootstrapping to evaluate the reliability of each discovered dependency in the database to avoid the risk of ascribing a causal role to a gene when not enough information is actually available to support the claim. Hartemink *et al.* [85] are tackling the underdetermination problem by tuning the unsupervised search of the most probable network structure. They leverage on established biological knowledge to select a small number of networks and then limit their comparisons to these networks only.

The use of BBNs in functional genomics is still in its infancy and the common problems of functional genomics data have still to be solved and much work is still left to render the technology fit for the task. Nonetheless, even these early efforts make clear the potential of these methods to dissect the inner structure of the regulatory circuitry of life.

5 Bio-Ontologies, Data Models, Nomenclature

Suppose that you have mastered normalization of microarray data sets, performed experiments to ensure reproducibility, and then gone on to acquire several large data sets and have now completed your clustering or classification experiments. Remarkably, after that preparation, you have what looks like a very nice set of results where the clusters of genes you found on one set of experiments have reproduced beautifully in another set of experiments that were designed to be identical. Furthermore, every time you slightly permute the conditions from those in the control or baseline condition, the sets of genes in the clusters seem to have changed in reproducible ways. And furthermore, the clusters appear to be compact and distinct so that you suspect that you have identified truly significant, functionally related genes.

If one steps back a little at this point, it becomes less obvious what the meaning is of all the effort that one has invested to date, because what one actually has generated is a list of microarray accession numbers. Even an experienced functional genomicist looking at this list of numbers will not have the vaguest idea of what it means. Some of these accession numbers cluster two or three genes, while other clusters include hundreds. At this point, the functional genomicist has very little alternative other than to look up the genes corresponding to each microarray accession number one by one and then peruse the literature regarding each one of the of the genes to determine

- whether that gene has a known function, and if so, in what class (*e.g.*, transcriptional factor, metabolic enzyme, structural protein, combinations of these, *etc.*);

- whether the genes found clustered together have been described in the literature as being functionally similar or related, or perhaps share promoter motifs, or a subset of the cluster are transcription factors for the rest of the cluster;

- whether homologs or orthologs have been found to be functionally related in any known physiological or pathological state;

- whether the resultant genes are known to be associated with the experimental conditions tested.

To the degree that the functional genomicist understands basic biology, this task is somewhat more tractable. A competent biologist will acquire in the course of his or her training a large framework of functional dependencies of different biological processes and the genetic machinery that underlies these dependencies. This kind of knowledge is unlikely to be present in any single journal article. Even a review

article presumes a large shared body of biological knowledge that will be opaque
to a researcher unfamiliar with a particular corner of the large space of biological
knowledge. Without this rich context, many insights that the analyses might suggest will be missed. Consequently, a biologist familiar with the specific biological
phenomenon being studied will be an invaluable collaborator in interpreting the
meanings of the results from analysis of these massively parallel data sets. However, even a competent biologist armed with a very efficient search engine of the
biomedical literature will find that determining whether the functional dependencies found in these clusters make any particular sense will take a very long time,
often more than any other part of the analytic methodology of functional genomics.

Consequently, after having undertaken this laborious, rote lookup process several
times, the following fantasy occurs to most genomicists: "Wouldn't it be nice if I
could look at the cluster and automatically see that one of the clusters contained
both genes coding for known transcriptional factors and genes coding for structural
proteins, and under two pathological conditions, those structural proteins were only
translated if the transcriptional factors in that same cluster were expressed at their
highest levels. Also, six researchers have developed mouse models in which four of
these genes in that cluster have been misexpressed, and those mouse models can
be obtained by filling out a form on a specified website..."

Enticing as this fantasy is, unfortunately, it does not represent the current state
of the art. However, there are many early efforts to achieve this goal. Some skeptics
will note that this is an "AI complete problem" [7] in the sense that reaching the
goal outlined in the above scenario would really be equivalent to significant progress
in the development of machine intelligence that can reason in a commonsensical
fashion about a wide range of heterogeneous knowledge. However, as we shall see
from some of the early efforts below, even a much more modest set of goals might
be quite useful for the purposes of the functional genomicist. The technological
solutions provided to achieve these goals fall into three increasingly circumscribed
tasks: development of bio-ontologies, development of common data models, and
development of common nomenclatures.

5.1   Ontologies

The term *ontology* is in its current usage substantially different from its original
philosophical meaning, which was the "study of the nature of being." However, as
most modern linguists will acknowledge, correct usage in the end is determined by
widespread current practice and currently, the term ontology encompasses the study

of formalized descriptions of all entities within the area of interest or research and all the relationships between those entities. In the discipline of functional genomics, bio-ontologies refer to organized and formalized systems of concepts of biological relevance in which the relationship between these concepts can be qualified. Before we proceed to some of the technical aspects of ontologies and examples of the more successful bio-ontologies, we list some notable efforts at creating usable bio-ontologies. First, we list here some additional motivations for the development of bio-ontologies.

- The need for a communal memory in the context of exponential growth in the amount of genomic information. That is, at the very least, we want to be able to share the biological knowledge that we are able to derive from our experiments, both to avoid duplication and to build on the experience of others.

- If we are to automate the process of sharing knowledge and use the prior inferences of functional dependencies, then the ontology must have sufficient formalization to enable a computational process that is both sound and efficient. As will be discussed briefly below (section 5.2), there is a natural tension between developing ontologies that are sufficiently expressive and those that are sound and efficiently computable.

- In order for a bio-ontology to be truly useful, it needs to be able to communicate the definition of terms in a reader- and context-independent way. This is particularly important in the domain of functional genomics because of the multiplicity of function of genes. For example, a gene's products will often have diverse functions within the same cell or even within the same subcellular compartment, and most often in different tissue types and under divergent environmental conditions. Stated in the converse, a gene's function will be highly determined by the subcellular context of its protein and the other coexpressed genes. For example, medium-chain dehydrogenase serves as a metabolic catalyst in one context, and in another it serves as a component of the structure of the eye lens (one of the crystallins) in some animals.

- An ontology should only represent the terms or relationships that are absolutely necessary (minimal ontological commitment). The lesser the ontological commitment, the greater the number of purposes for which a particular ontology can be used. For instance, consider a bio-ontology that is specifically designed to capture the structural protein motifs associated with a gene and to capture the potential sites of interaction between proteins. This ontology may be extremely useful in the screening of genes that might be coding for proteins that are discovered to physically interact, but it may be less useful in determining whether or not a particular protein is a known transcriptional factor for a particular gene. This latter information might best be represented in a different formalism than the protein structure representation.

- There is a significant need for *organism-specific* ontologies, as genes do not always have the same function across different species. A gene in the human genome may have a single homolog in yeast, but the way in which it is regulated, the biological systems in which it participates, and the nature of the additional structural complexity of its protein in the human may be obscure without a species-specific bio-ontology.

- There also needs to be *cross-organism* ontologies because many genes have identical or analogous function across organisms. If a finding is made in a yeast experiment, *e.g.*, then the ontology should inform the functional genomicist of the potential for finding a human gene with the same function, if one is found with similar sequence homology. Conversely, a gene in one species may have similar function to that of another nonhomologous gene in another species. Understanding the parallels in the regulation and function of different genes across species may provide important biological insights.

5.1.1   Bio-ontology projects

We will now cover a few examples of successful bio-ontologies. For a more comprehensive and current compendium of ontologies such as these, the reader is referred to issue no. 1 of *Nucleic Acids Research*, of any recent year.

The Gene Ontology Consortium   The Gene Ontology (GO) consortium is one of the more high-profile efforts that has developed in response to the pressing need for a unifying framework. As summarized by Ashburner *et al.* [13], the goal of the GO consortium is to provide a controlled vocabulary for the description of three independent ontologies:

1. *A description of molecular function.* This describes the biochemical activity of the entity such as whether it is a transcriptional factor, a transporter, or an enzyme, without making any further commitments as to where the functions occur or whether the function occurs as part of a larger biochemical process.

2. *The cellular component within which gene products are located.* This provides a localization of a gene product's molecular function, such as the ribosome, the nu-

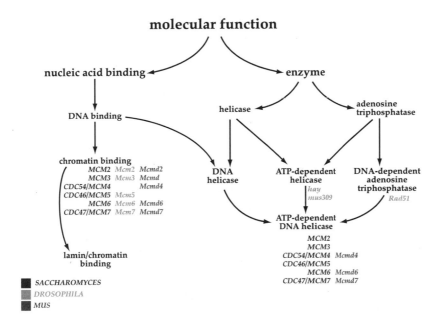

Figure 5.1
The Molecular Function Ontology of GO. (From Ashburner *et al.* [13].)

cleosome, or mitochondrion. Localization can help determine whether a purported function could occur through direct physical interaction between gene products or as a result of an indirect mechanism.

3. *The biological process implemented by the gene products.* Biological process refers to a higher-order process, such as pyrimidime metabolism, protein translation, or signal transduction.

The goal is to allow queries across databases using GO terms providing linkage of biological information within and across species. The three ontology domains are presented in figures 5.1, 5.2, and 5.3.

These three ontologies are a result of a practical consensus of the kinds of distinctions that are important at present in functional genomics. If the GO consortium efforts were to populate this form of annotation across all genes for all sequenced organisms, this would be an invaluable resource even though the ontology is of so little commitment that it may not enable several categories of useful inferences. Examples of the latter include causal transitivity, temporal ordering, assertion of

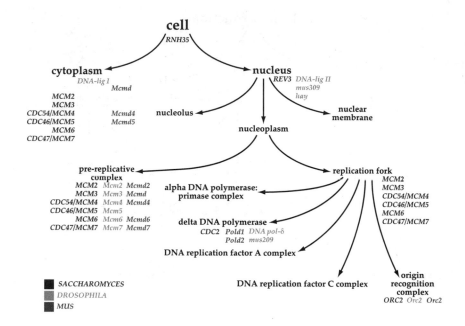

Figure 5.2
The Cellular Component Ontology of GO. (From Ashburner *et al.* [13].)

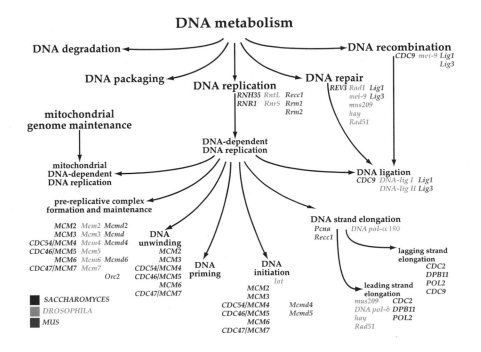

Figure 5.3
The Biological Process Ontology of GO. (From Ashburner *et al.* [13].)

disjoint subconcepts, and many more. Representations which might support these more elaborate inferences are discussed in section 5.1.2. Again, there are some significant trade-offs in using ontologies of much more ontological commitment and expressivity (see section 5.2). Despite the limitations of the simple ontologies of GO, early efforts to annotate the organismal genomes using this infrastructure appear very promising, notably the GO annotations in the LocusLink site of the National Center for Biotechnology Information (NCBI).[1]

Kyoto Encyclopedia of Genes and Genomes   One of the more comprehensive attempts to describe the molecular and functional biological knowledge (all the way down to the detail of interacting molecules) for all known genes is the Kyoto Encyclopedia of Genes and Genomes (KEGG).[2] Rather than generating annotations per gene, KEGG attempts to curate and elaborate on metabolic pathways, large molecular assemblies, and regulatory pathways. These larger abstractions are invaluable to biological researchers and allow the rapid identification of how a gene fits into cellular physiology. The KEGG ontology has five components:

1. *Pathways* with a corresponding graphic pathway map

2. *Ortholog groups*, which represent highly conserved functional groups annotated by the genes belonging to these groups across species

3. *Molecular catalogs* providing functional annotation of proteins, RNA, and small molecules

4. *Genome maps*, which mirror much of the positional information at the NCBI

5. *Gene catalogs* which have similar content to the Molecular Function Ontology of GO

The richness and utility of the KEGG ontology can be seen in figure 5.4 which provides a snapshot of one pathway maintained in its Pathways ontology. It documents our increasingly complex state of knowledge about the genetic regulation of programmed cell death—apoptosis.

5.1.2   Advanced knowledge representation systems for bio-ontology

The AI community has developed a set of tool kits for large and highly portable ontologies over the past two decades. It has not escaped that community's notice

---

[1]http://www.ncbi.nlm.nih.gov/LocusLink/.
[2]http://www.genome.ad.jp/kegg/kegg2.html.

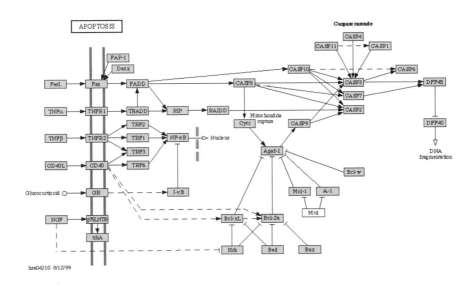

Figure 5.4
The KEGG representation for apoptosis.

that the bio-ontologies developed to date do not exceed the representational capabilities of extant knowledge representation languages. In particular, languages such as Ontolingua [142] and the Ontology Mark-up Language/Conceptual Knowledge Mark-up Language (OML/CKML) [131] have been evaluated for their ability to represent biological knowledge and specifically functional genomic relationships. Ontolingua is a spin-off of the Knowledge Interchange Format (KIF) and is a result of research funded by the Advanced Defense Research Project Agency Knowledge Sharing Effort [137] targeted to support knowledge re-use in ontology building efforts, a target shared by the bioinformatics community. The OML/CKML effort has been led by Washington State University. These two representation languages have similar capabilities as described in a report by McEntire *et al.* [131], although at this time, the latter has a more compatible syntax with other meta-data languages through its use of eXtensible Mark-up Language (XML). The reader is referred to [131] for a description of the issues that had to be addressed when representing existing genomic knowledge bases.

Another knowledge representation system that has been useful in the bio-medical domain is PROTÉGÉ, designed by Mark Musen at Stanford. In practice, none of these more advanced knowledge representations has had nearly the impact and usage of the simple ontologies such as that of GO or Proteome.[3] Part of the reason is that the need for any kind of functional annotation is so dire that the capabilities of the advanced knowledge representations (*e.g.*, automated inferences) are not a priority and therefore unnecessary overhead when applying them over tens of thousands of genes. It is also the case that even though these advanced representations have significant expressive power, they remain fundamentally limited as described below. We wish to emphasize this point because it is only through the implementation of bio-ontologies with greater expressive power that many of the properties desired for such ontologies can be achieved. The alternative is the creation of a multiplicity of special-purpose, limited expressivity representations that can only be made interoperable with significant effort and without guarantees of semantic or computational soundness.

5.2   Expressivity versus Computability

Since the 1980s, it has been recognized that there is a natural trade-off between the expressivity of a representation used for an ontology and its computability. In an illuminating paper by Haimowitz *et al.* [82], an evaluation was made of the adequacy of a formal description language—NIKL—to represent medical concepts used in a physiological diagnostic program. The authors of this paper described the requirements of a knowledge representation language to include the following desiderata:

- The system should have precise semantics, either based on the semantics of first order predicate calculus, or having at least comparable precision.

- The knowledge representation should automatically provide logical inferences. This is what sets a knowledge representation apart from a conventional database. It can answer questions that go beyond what was explicitly told to the system.

- It should include some form of taxonomical inheritance of characteristics by more specialized concepts from more general ones. Variations on the inheritance mechanism include whether it is a pure tree (only one parent or generalization for each concept) or more general graph (multiple potential generalizations of each concept). For example, an ontology could represent the fact that a protein can be both part of a signal transduction pathway and a transcriptional factor. Some inheritance mechanisms allow for exceptions and others do not (*e.g.*, although most RNA molecules code for proteins, some do not.).

---

[3]http://www.proteome.com.

- The inference mechanism should be sound and complete; *i.e.*, it should derive no false conclusions from true knowledge and guarantee that all true conclusions within the class automatically promised will in fact be made.

- Whatever automatic inference is provided by the representation system itself should be efficient.

- Because of the requirements of soundness, completeness, and efficiency, the expressive power (the kinds of things that can be said with it and inferred in the knowledge representation) should be limited to avoid the representation of undecidable or "NP-hard" concepts.

These desiderata apply to all the existing bio-ontologies and, therefore, the limitations that Haimowitz *et al.* found in their evaluation are likely to hold in the construction of bio-ontologies. Among the troublesome aspects of applying NIKL to the medical domain, Haimowitz *et al.* found the representation to be inadequate to capture spatial relationships so that, *e.g.*, the transitivity of the part-whole or contained relationships was not automatically preserved. Furthermore, causation has a variety of flavors (*e.g.*, a gene may be permissive for a particular process, or a gene product may be necessary but not sufficient for that process, or a gene product might be necessary and sufficient) and these could not easily or accurately be represented in NIKL. Also, the transitivity relationships that would then be expected of a causal representation were not suitably automatically inferred. Similarly, in the temporal representation of the different phases of a process, the dependencies of overlapping phases of a process were not adequately captured. These limitations would certainly create problems for representing large numbers of cellular processes germane to functional genomics.

Subsequently, several special-purpose representations were developed to create ontologies that met the list of desiderata listed above. These specialized knowledge representations pertained to temporal reasoning or spatial reasoning or causality. A few of them had limited capabilities across all of these aspects. None of the current bio-ontologies are represented in languages as rich as these efforts dating back to the 1980s. The implication is that, for our foreseeable future, the expressivity that can be reasonably expected from bio-ontological representation languages is going to be limited and we will not be able to express all the kinds of relationships that a biologist wishes to assert. Nonetheless, for the reasons noted at the beginning of this section, the utility of having even a relatively inexpressive set of attributes for concepts associated with genes and gene interactions would be immense. The alternative is manual review of the unstructured literature, and that is unlikely to be a scalable solution without further efforts in natural language-based information retrieval techniques. As we have stated before, notwithstanding their limitations, these bio-ontologies are likely to grow in size and in use. Indeed there are several commercial efforts now that have implemented very low ontological commitment ontologies such as that of Proteome in which very simple ontologies have been populated by hundreds of knowledge workers to supply the biotechnology industry with annotations for the genome. Similar efforts are underway by the other gene discovery groups and companies in this area such as Celera, Curagen, and Rosetta.

A comparison of the various ontologies mentioned in this section is shown in table 5.1.

5.3   Ontology versus Data Model versus Nomenclature

In the sections below, we describe the nomenclatures and data models that are used to represent genomic data. Often, however, data models and nomenclatures are described as being ontologies. In some very abstract sense, a data model is an ontology with very little ontological commitment (and little in the way of automated inferential mechanisms), and a nomenclature has even less. A *nomenclature* simply refers to a method of naming or identifying concepts. From the ontological perspective, a nomenclature allows the representation by name of the concepts in an ontology (whether implicit or explicit) without describing the relationships between these various concepts.

A *data model* defines a hierarchy of classes, attributes and relationships of those attributes without providing the expressivity based on first-order logic. It also does not have the other inferential capabilities that we have mentioned in the desiderata of section 5.2 and the description of the NIKL effort of Haimowitz *et al.* For instance, a data model describing the components of a microarray (see section 5.4) might be useful for creating a shared structure for storing the results of a microarray experiment. This data structure does confer the capability to provide even simple inferential relationships such as negation or disjunction. However, a data model would be awkward to state that a gene can code for a transfer RNA or an rRNA or a protein but only two of these three classes. Certainly, elaborate data structures can be developed to capture representational detail, but for every disjunction or exception that the ontology architect wishes to add, the complexity of the data model will grow rapidly. That is, while the data model is necessary to provide a structure in which to maintain the properties of the concepts, it is no substitute for the descriptive power and inferential power of the kind of ontologies previously

Ontology	Defines a nomenclature for genomics/bioinformatics	Specialization	Expressivity	Interoperability / Integration	Human expertise required
Ontolingua	Multipurpose tool; no nomenclature included or recommended. User community can define the nomenclature of choice.	Support of multiple inheritance allows representation with several add-on packages for specialized domains (e.g., temporal and spatial reasoning). Can assert disjointness, negation, etc. This expressivity unsurprisingly is coupled with minimal reasoning support.	Broad and richly typed representation with several add-on packages for specialized domains (e.g., temporal and spatial reasoning). Can assert disjointness, negation, etc. This expressivity unsurprisingly is coupled with minimal reasoning support.	Several tools from within the Artificial Intelligence community are interoperable. Less so with standard WWW tools.	Considerable knowledge acquisition required for proper use.
GO	Specifies a controlled vocabulary for GO descriptions	Limited specialization without formal inheritance mechanisms.	Limited expressivity (no causal transitivity, temporal ordering, assertion of disjoint subconcepts).	Tightly integrated into the suite of databases and tools of the NCBI.	Minimal ontological commitment and restricted expressivity enable easy adoption and use.
KEGG	Specifies standard nomenclature for each entity (e.g., EC for enzymes).	Very limited form of specialization without inheritance.	Limited expressivity (no assertion of disjoint subconcepts). However, the extensive use of binary relations allows primitive interference of causal transitivity and representation of pathways.	Integrated with other pathway representations (e.g., WIT[a]).	Minimal ontological commitment and restricted expressivity enable easy adoption and use.
OML/CKML	Multipurpose tool; no nomenclature included or recommended. User community can define the nomenclature of choice.	Supports multiple inheritance and automated subsumption, i.e., inference from the logical constraints defined in the term of concept C' logically imply those of the more general concept C.	Missing some of Ontolingua's constructs (e.g., collections) but relatively high expressivity.	Uses XML as syntax and therefore easily interoperates at least at the syntactic level with several e-commerce/WWW tools. As yet, though, few instances of integration with other tools.	Significant experience required but XML syntax brings it quicker to wider community of developers. However, still a work in progress.

Table 5.1
Overview of representative ontology technologies. "Consumer report" of instances of some of the dominant ontologies used to represent genomic knowledge. Some of the instances are general purpose ontology building tools (e.g., Ontolingua) and others are the sole instance of a particular representational technology (e.g., KEGG).

---

[a]http://wit.mcs.anl.gov/WIT/

mentioned. This may appear obvious, but at the time of this writing there are
several efforts which claim to be ontologies, but which are in fact no more than
glorified nomenclatures and data models.

5.3.1 Exploiting the explicit and implicit ontologies of the biomedical literature

Instead of devising new ontologies specifically for use in functional genomics, others
have taken advantage of the existing very low ontological commitment representations used to maintain the biomedical literature. The terminology structures of
MEDLINE, the Medical Subject Headings (MeSH), are an outgrowth of the original paper-based *Index Medicus*. As illustrated in figure 5.5 below, MEDLINE's key
words are organized into major headings within a few fixed hierarchies or trees (*e.g.*,
physical findings, diseases, chemicals). This limited taxonomical structure makes it
evident that MeSH has limited expressive capabilities. It has the immense benefit
of having been carefully curated by extremely well trained librarians over decades,
and therefore it represents an invaluable depository of annotation of biomedical
knowledge. It includes approximately 20,000 major headings (and many more synonyms corresponding to the concepts represented by these headings) and 105,000
chemical names. Each article encoded in MEDLINE (and therefore accessible via
PUBMED[4]) has on average 10 MeSH annotations.

The MeSH nomenclature could be taken advantage of in functional genomics. For
example, the work of Dan Masys *et al.* in the HAPI [127] project processes microarray
data and clusters from functional genomics analyses. The genes are looked up and
the corresponding MeSH terms are found in databases maintained by the NCBI (see
the nomenclature section 5.5 below). The MEDLINE structure associated with each
publication is then used to categorize the genes found in the clusters, as shown
in figure 5.6.[5] That is, in addition to finding the publications corresponding to
each gene, the MeSH hierarchy permits the identification of broader functional and
pathological categories. In the instance of the clusters of genes characterizing acute
myelogenous leukemia and acute lymphocytic leukemia [78], this demonstrates that
the genes predictive of these leukemias are also known to be involved in polycystic
kidney disease, inherited immunodeficiencies, and multiple sclerosis. Many of these
relationships were only obtainable through the MeSH hierarchy and not directly by
searching using only the gene names.

---

[4]http://www.ncbi.nlm.nih.gov.

[5]A similar functionality is also provided by the PubGene program at http://www.pubgene.org.

MeSH Tree Structures - 2001

1. ⊞ **Anatomy [A]**
2. ⊞ **Organisms [B]**
3. ⊟ **Diseases [C]**
    - ○ **Bacterial Infections and Mycoses [C01]** +
    - ○ **Virus Diseases [C02]** +
    - ○ **Parasitic Diseases [C03]** +
    - ○ **Neoplasms [C04]** +
    - ○ **Musculoskeletal Diseases [C05]** +
    - ○ **Digestive System Diseases [C06]** +
    - ○ **Stomatognathic Diseases [C07]** +
    - ○ **Respiratory Tract Diseases [C08]** +
    - ○ **Otorhinolaryngologic Diseases [C09]** +
    - ○ **Nervous System Diseases [C10]** +
    - ○ **Eye Diseases [C11]** +
    - ○ **Urologic and Male Genital Diseases [C12]** +
    - ○ **Female Genital Diseases and Pregnancy Complications [C13]** +
    - ○ **Cardiovascular Diseases [C14]** +
    - ○ **Hemic and Lymphatic Diseases [C15]** +
    - ○ **Neonatal Diseases and Abnormalities [C16]** +
    - ○ **Skin and Connective Tissue Diseases [C17]** +
    - ○ **Nutritional and Metabolic Diseases [C18]** +
    - ○ **Endocrine Diseases [C19]** +
    - ○ **Immunologic Diseases [C20]** +
    - ○ **Disorders of Environmental Origin [C21]** +
    - ○ **Animal Diseases [C22]** +
    - ○ **Pathological Conditions, Signs and Symptoms [C23]** +
4. ⊞ **Chemicals and Drugs [D]**
5. ⊞ **Analytical, Diagnostic and Therapeutic Techniques and Equipment [E]**
6. ⊞ **Psychiatry and Psychology [F]**
7. ⊞ **Biological Sciences [G]**
8. ⊞ **Physical Sciences [H]**
9. ⊞ **Anthropology, Education, Sociology and Social Phenomena [I]**
10. ⊞ **Technology and Food and Beverages [J]**
11. ⊞ **Humanities [K]**
12. ⊞ **Information Science [L]**
13. ⊞ **Persons [M]**
14. ⊞ **Health Care [N]**
15. ⊞ **Geographic Locations [Z]**

Figure 5.5
The MeSH terminology hierarchy. A portion of the MeSH terminology hierarchy starting at the top level and showing the first level of detail at the disease subhierarchy. The entire hierarchy may be browsed at http://www.nlm.nih.gov/mesh/MBrowser.html.

**Diseases Associated with
ALL-predictive genes**

```
Virus Diseases (1) {>.3}
Neoplasms (5) {>.3}
   Neoplasms by Histologic Type (4) {>.6}
      Leukemia (4) {<.001}
         Leukemia, Lymphocytic (3) {<.001}
            Leukemia, B-Cell (1) {<.001}
               Leukemia, B-Cell, Acute (1) {<.001}
            Leukemia, Lymphocytic, Acute (1) {<.001}
               Leukemia, B-Cell, Acute (1) {<.001}
            Leukemia, T-Cell (1) {<.001}
      Precancerous Conditions (1) {<.001}
         Preleukemia (1) {<.001}
Nervous System Diseases (2) {>.3}
   Autoimmune Diseases of the Nervous System (1) {<.001}
      Demyelinating Autoimmune Diseases, CNS (1) {<.001}
         Multiple Sclerosis (1) {<.001}
   Demyelinating Diseases (1) {<.001}
      Demyelinating Autoimmune Diseases, CNS (1) {<.001}
         Multiple Sclerosis (1) {<.001}
Female Genital Diseases and Pregnancy Complications
   (1){>.6}
   Genital Diseases, Female (1) {>.6}
      Infertility (1) {<.001}
         Infertility, Female (1) {<.001}
Hemic and Lymphatic Diseases (1) {>.6}
   Hematologic Diseases (1) {>.6}
      Preleukemia (1) {<.001}
Neonatal Diseases and Abnormalities (3) {>.3}
   Hereditary Diseases (2) {>.13}
      Werner Syndrome (1) {<.001}
   Infant, Newborn, Diseases (1) {<.001}
      Severe Combined Immunodeficiency (1) {<.001}
Immunologic Diseases (4) {<.01}
   Autoimmune Diseases (1) {>.13}
      Autoimmune Diseases of the Nervous System (1){<.001}
         Demyelinating Autoimmune Diseases, CNS (1){<.001}
            Multiple Sclerosis (1) {<.001}
   Immunologic Deficiency Syndromes (3) {<.001}
      Common Variable Immunodeficiency (1) {<.001}
      Severe Combined Immunodeficiency (1) {<.001}
Pathological Conditions, Signs and Symptoms (1) {>.3}
   Pathologic Processes (1) {>.13}
      Disease Attributes (1) {<.001}
         Acute Disease (1) {<.001}
```

Figure 5.6
Transforming a cluster of genes into an annotated structure automatically using MEDLINE.
Summary of concept hierarchy matches with the MeSH hierarchy terms for genes described by
Golub *et al.* [78] that fell into the acute lymphoblastic leukemia (ALL) cluster.

Alternatively, the textual content of the biomedical literature itself has an implicit (if muddy) ontological structure which can be exploited. That this can be done without solving the natural language challenge is illustrated by the work of Altman and Raychaudhuri [6]. These authors represented each publication as a word vector of the titles and abstracts, with each vector encoding the frequency of those words for those articles. Using these vectors they were able to apply the same clustering techniques[6] used to find the proximity of gene expression patterns to find how the publications clustered. It may be that in the future, combinations of simple ontologies and a statistical analysis of the content of publications will allow efficient mining of the immense biomedical literature for genomicists seeking to obtain the meaning of the gene or protein patterns they have observed.

5.4   Data Model Introduction

Data models are the logical structures used to represent a collection of entities and their underlying one-to-one, one-to-many, and many-to-many relationships. The main motivations for creating data models is usually to be able to implement them within a database management system, usually as a relational database management system (*e.g.*, Oracle, SQL Server, Sybase, MySQL, *etc.*). As the relational calculus [48] is based on the first-order predicate calculus, it is possible, although unrewarding, to express a wide range of formal relationships within the data model. The reasons for the unwieldy nature of such an effort are articulated in the prior section. Presently, in the domain of bioinformatics and functional genomics, the most relevant heuristic distinction between bio-ontologies and data models is that bio-ontologies are usually developed in order to provide a shareable framework to describe attributes of genes, whereas data models at this time are mostly developed to store the measurements of genomic systems. For DNA sequences, this corresponds to the GenBank data model and for the domain of microarrays, it corresponds to a way of organizing the pertinent values obtained on measurement of a microarray hybridization experiment.

Among the goals often stated for the development of standardized microarray data models is the capability to allow researchers across multiple laboratories to be able to take advantage of the precious RNA samples obtained from various sources and hybridized to microarrays in several, initially unrelated, experiments. Ideally, these researchers should be able to perform aggregate analyses across all these various experiments independent of the provenance or manufacture of each

---

[6]With the word frequency used in the same way as a gene expression value.

of the microarray systems. In this way, we may create a shared set of repositories of expression data that might reveal subtle gene interactions across a larger sample than would be observable within a single institution's data set.

Unfortunately, expression measurements are human artifacts rather than ground "truths" in biology.[7] A consequence of this fact is that that there are many such artifacts. There are several microarray measurement systems in use, and many more under development, such that convergence upon a particular data model involves the arduous task of obtaining a sufficiently general representation of all the measurement types across different microarray systems, while at the same time providing sufficient specificity for each measurement system so as to avoid loss of detail of a particular study. Unfortunately, a large part of this specificity reflects differences in the measurement techniques of a particular microarray platform that result in nonlinear relationships with the results from other platforms. The implication of this is that even if the data models are shared across microarray platforms, the results encoded in these shared data models may not be directly comparable.

Let us be a little bit more explicit regarding the challenge of a standardized data model for microarray experiments results. Unlike DNA sequences, which are by nature digital and whose interpretation should be independent of the modality by which the sequence was acquired, expression data are analog signals and their value and interpretation are very much dependent on the technology used to obtain that value. Consider just two of the technologies used to measure expression: photolithographically constructed oligonucleotide microarrays with several probes per gene hybridized against a single color-stained sample, and robotically spotted microarrays with one probe per gene hybridized against two samples each colored with a different dye. In the first case, the expression level is reported as a function of a trimmed mean of intensities, each probing for a particular part of a gene, and the "average difference" between intensities of oligonucleotides that perfectly match against that gene part, and oligonucleotides that have a central base mismatch with that same part. In the second case, the gene expression level is measured as a ratio of hybridization of one specific cDNA from one sample against the cDNA from another sample. Although there is some relationship between the two measurements, it is not clear or empirically determined what that relationship is. For example, two-dye microarray measurements are directly related to the second or control sample used, whereas single-dye oligonucleotide microarray measurements are in theory an absolute expression level.

---

[7]That is, this compares with the case of DNA sequence, where the way to represent the ordering of a particular sequence of nucleotides is, at least in theory, independent of the measurement system.

Sample fields

- **Experiment type**: Single channel (absolute value) such as radioactive labeled filters, or dual channel types such as a red/green labeled microarray.
- **Sample title**: Try to choose a short title which will be specific to your sample -- preferably greater than 8 and less than 20 characters. The title must be unique over all your previously accessioned GEO samples. You will be notified if such a name clash occurs when you try to submit, and then will be given a chance to change your sample title.
  Text is required in this field.
- **Lot/batch**: Any manufacturer defined lot and/or batch numbers for the platform used to derive this sample's data.
- **mRNA source**: It is best to use a short list of words to describe the source of the mRNA used to derive this sample's data. Do not list organisms in this field.
  Text is required in this field.
- **Ch1 mRNA source**: It is best to use a short list of words to describe the source of the mRNA used to derive this sample's data. Do not list organisms in this field.
  Text is required in this field.
- **Ch2 mRNA source**: It is best to use a short list of words to describe the source of the mRNA used to derive this sample's data. Do not list organisms in this field.
  Text is required in this field.
- **PubMed id**: PubMed id (PMID) references a publication which, perhaps, further describes your sample. A PubMed id is a numeric value which you may obtain from the PubMed record.
- **Web link**: World Wide Web link to reference a webpage which, perhaps, further describes your sample.

- **Keywords**: One or more terms, as a comma-delimited list, which you think might be useful terms in queries.
- **Image file**: Use the Browse button to the right to select a local file to be uploaded. The purpose of this image is to serve as a qualitative reference of your experiment. It is not meant to be image-analyzable. In order to be accepted into GEO, the image size must be less than 100 kilobytes, and must be in JPEG or GIF format.
- **Total # tags**: A whole, non-zero number is required. The reciprocol of this number is used for SAGE library normalization.
- **Norm. factor**: This number is meant to be multiplied to all of the background-corrected scalar values extracted from each platform feature referred to in the data table.
  A number is required.
- **Protocol**: Select the anchoring enzyme used in your SAGE protocol, e.g., NlaIII or Sau3A. Protocol selection in SAGE libraries replaces platform selection.
  A selection is required.
- **Platform used**: Submission of non-SAGE sample data requires explicit reference to a valid GEO platform accession number. Therefore, platform information must be submitted before sample data derived from that platform.
  A selection is required.
- **Data file**: Use the Browse button to the right to select a local file to be uploaded. This file should meet the validation criteria set for the platform type relevant to your sample.
  A content bearing, valid data file is required.

Figure 5.7
A subset of the definition of *sample* in the GEO. At the time of this writing, the sample definition in the GEO does not appear to support the full complexity of oligonucleotide high-density microarrays.

As a result, there are at least as many microarray expression data models available as there are microarray expression technologies and, in fact, considerably more. At the time of this writing, there has yet to be any consolidation of these different competing standards into a single one, although this is likely to happen in the near future. However, in the absence of such a consensus, we point out the most promising and successful of these efforts. Not all these efforts in data modeling are sufficiently general at the time of this writing to encompass all the current microarray gene profiling technologies. We will note this as we describe the current major candidates for standardized data models.

Gene Expression Omnibus (GEO)   The NCBI has been developing the GEO data model to support the construction of a gene expression repository with the same inspiration and goals that were originally targeted for the GenBank repository for DNA sequence. The goal of GEO is to support spotted microarray technology, oligonucleotide microarray technology, hybridization filters, and Serial Analysis of Gene Expression (SAGE) data. The entities used in GEO are shown in figure 5.7 and can be accessed on-line.[8] Examination of this figure reveals that, currently, Affymetrix style files are poorly supported, whereas the two-dye spotted microarray data type is better supported.

---

[8] http://www.ncbi.nlm.nih.gov/geo/info/fields.pgi.

The Gene Expression Mark-Up Language (GEML)   GEML syntax is based on XML. GEML was originally developed by Rosetta Inpharmatics, and subsequently, several other public and private institutions have become involved, such as Agilent Technologies and Nature America, Inc. Details on GEML can be found at http://www.geml.org.

GEML starts with an XML format, as this is the most widely used meta-data syntax developed. XML is maintained by the World Wide Web consortium to define general-purpose mark-up languages, and is now used widely as the interchange format for a variety of electronic commerce functions and scientific interchange efforts. There is a multiplicity of tools for generating, parsing, and displaying XML content and consequently GEML leverages this existing effort. The goal of GEML is illustrated in figure 5.8 (taken from their website). At the top level are the various types of microarray measurement systems and in the middle is the GEML *lingua franca* into which all the values measurement systems are translated. Below the GEML layer are shown several applications that work with microarray data as abstracted into a standardized format in GEML. This architecture is similar to many other three-tiered architectures where there is a common data model abstracting away the heterogeneity and complexity of the various databases that it accesses. In theory, a programmer for microarray analysis software would only have to understand the GEML Document Type Definition (DTD) to write applications that worked with the results of measurements on all microarray platforms.

The GEML specification takes advantage of the DTD file. A DTD is the actual specification of the formal data model that any XML message that hews to that DTD must follow. DTD specifications are the World Wide Web consortium's prescribed mechanism for defining application-specific data models, and many industries (*e.g.*, finance and imaging) are in the process of defining a consensus DTD for their own applications.[9] GEML specifies two DTDs: a pattern DTD, which describes the genes reported on a chip layout, and the profile DTD, which contains expression data, treatment, and hybridization information. That is, the latter DTD describes the experimental conditions and the resultant values. figure 5.9 shows the pattern DTD that has been quoted verbatim from the GEML site. Examination of this DTD reveals that the encoding of the experimental results of single or two-dye stained samples hybridized against cDNA microarrays would be fairly straightforward, whereas encoding the 20 probe values that constitute a single probe set on

---

[9]In the latest iteration of specifications for XML document specifications, the XML schema has superseded the DTD, although there is sufficient overlap to allow DTDs to be easily converted into XML schemas.

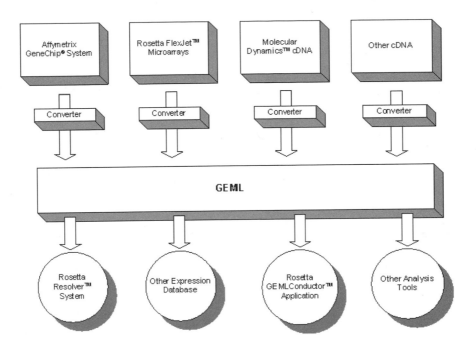

Figure 5.8
The abstraction layers of the Gene Expression Mark-up Language (GEML).

an oligonucleotide microarray might be challenging and require some contortions in the way the data is squeezed into those existing attribute lists.

Microarray Gene Expression Database Group (MGED)   MGED was developed in recognition of the need for standardization of microarray data models. It is an open discussion group that originally held its meetings in the United Kingdom, and subsequently has held two international meetings.[10] The most notable output of the MGED group, which is constituted of multiple private and public institutions, is the Microarray Mark-up Language (MAML). The MAML DTD is shown in figure 5.10. The MAML DTD is sufficiently comprehensive that even at the time of this writing, it appears to cover all the currently available microarray technologies without requiring each site implementing the MAML database to put their own data through too many contortions in order to adhere to the data model. In particular, the abstract way in which a composite element has been defined in this DTD allows

---

[10]http://www.mged.org/.

```
<!-- GEMLPattern.dtd   -->

<!ELEMENT project   (pattern*,              <!ELEMENT gene    (accession*,
                     printing*,                                alias*,
                     other*)>                                  other*)>
<!--Project = group of patterns            <!--Gene = what the reporter
                  and/or printings-->                         is reporting on-->
<!ATTLIST project      name CDATA #IMPLIED  <!ATTLIST gene   primary_name CDATA #REQUIRED
                         id CDATA #IMPLIED                 systematic_name CDATA #REQUIRED
                       date CDATA #IMPLIED                         species CDATA #IMPLIED
                         by CDATA #IMPLIED                      chromosome CDATA #IMPLIED
                    company CDATA #IMPLIED >               map_position CDATA #IMPLIED
<!ELEMENT pattern    (reporter+,                            description CDATA #IMPLIED>
                      other*)>              <!ELEMENT accession (other*)>
<!--Pattern = collection of                <!ATTLIST accession  database CDATA #REQUIRED
                   one or more features-->                       id CDATA #REQUIRED>
<!ATTLIST pattern         name CDATA #IMPLIED  <!ELEMENT alias  (other*)>
                       type_id CDATA #IMPLIED  <!ATTLIST alias      name CDATA #REQUIRED>
              species_database CDATA #IMPLIED  <!ELEMENT position (other*)>
                   description CDATA #IMPLIED  <!ATTLIST position      x CDATA #REQUIRED
                        access CDATA #IMPLIED                          y CDATA #REQUIRED
                         owner CDATA #IMPLIED>                     units CDATA #REQUIRED>
<!ELEMENT reporter   (feature+,             <!ELEMENT pen      (other*)>
                      gene?,                <!ATTLIST pen            x CDATA #REQUIRED
                      other*)>                                      y CDATA #REQUIRED
<!ATTLIST reporter           name CDATA #REQUIRED               units CDATA #REQUIRED>
                  systematic_name CDATA #REQUIRED  <!ELEMENT printing   (chip+,
                      accession CDATA #IMPLIED                           other*)>
                       deletion CDATA "false"  <!ATTLIST printing      date CDATA #IMPLIED
                   control_type CDATA "false"                        printer CDATA #IMPLIED
                      fail_type CDATA #IMPLIED                          type CDATA #IMPLIED
               active_sequence CDATA #IMPLIED                   pattern_name CDATA #IMPLIED
               linker_sequence CDATA #IMPLIED                run_description CDATA #IMPLIED>
              primer1_sequence CDATA #IMPLIED  <!ELEMENT chip     (other*)>
              primer2_sequence CDATA #IMPLIED  <!ATTLIST chip       barcode CDATA #REQUIRED>
                   start_coord CDATA #IMPLIED  <!ELEMENT other    (other*)>
               mismatch_count CDATA #IMPLIED>  <!ATTLIST other        name CDATA #REQUIRED
  <!--Reporter = measures expression                                value CDATA #REQUIRED>
                         of a gene-->
<!ELEMENT feature    (position,
                      pen?,
                      other*)>
  <!--Feature = location of a
                   reporter for a gene-->
<!ATTLIST feature       number CDATA #IMPLIED
             ctrl_for_feat_num CDATA #IMPLIED>
```

Figure 5.9
The GEML pattern document type description.

for the representation of a wide range of physicochemical means for assessing gene expression.

Data model or database	URL
Affymetrix Analysis Data Model (previously Genetic Analysis Technology Consortium)	`http://www.affymetrix.com/support/aadm/aadm.html`
Another Microarray Database	`http://www.microarrays.org/software.html`
ArrayDB	`http://genome.nhgri.nih.gov/arraydb`
ArrayExpress	`http://www.ebi.ac.uk/arrayexpress/Design/design.html`
ExpressDB	`http://twod.med.harvard.edu/ExpressDB`
GeneX MicroArray Database	`http://www.ncgr.org/research/genex` `http://pompous.swmed.edu/exptbio/microarrays/mad`
Stanford Microarray Database	`http://www.genome-www4.standford.edu/MicroArray/SMD`

Table 5.2
List of freely available data models and databases for microarrays.

Attempting a consensus model Obviously, if there is such a wide variety of standard microarray models, there is no standard. Without a single standard, combining expression data across microarray experiments is arduous, if not impossible. Fortunately, the Object Management Group (OMG) appears to be creating a common ground where proponents of these various data models will be able to begin to reconcile their various proposals and create a consensus data model. Although this consensus data model has not yet been finalized, the OMG has already incorporated some of the proposals that we listed previously. As the OMG has successfully created agreed-upon standardized data models in many different application domains outside biology, there is some reason to hope that they will be similarly successful in this domain. For the latest information on the OMG efforts regarding a standardized microarray data model, we will refer the reader to the following website: `http://www.geml.org/omg.htm`.

```
<!ELEMENT creation_info    (contact,          <!ELEMENT array_platform  (array_def |
                            software*,                                    reference)          >
                            hardware*)    >   <!ATTLIST array_platform  id ID   #REQUIRED>
<!ATTLIST creation_info    date DATE  #REQUIRED>
                                               <!ELEMENT array_def       (creation,
<!ELEMENT contact          (parameter*)    >                             description?,
                                                                         comment*,
<!ATTLIST contact    last_name CDATA #REQUIRED                           reference*,
                    first_name CDATA #IMPLIED                            treatment*,
                   middle_name CDATA #IMPLIED                            parameter*,
                           lab CDATA #IMPLIED                            seq_feature*)      >
                    department CDATA #IMPLIED
                   institution CDATA #REQUIRED <!ATTLIST array_def       name  CDATA #REQUIRED
                        street CDATA #IMPLIED            surface_type  (non-absorptive |
                          city CDATA #IMPLIED                          absorptive |
                province_state CDATA #IMPLIED                          other)#REQUIRED
                       country CDATA #REQUIRED       other_surface_type  CDATA #IMPLIED
                postal_zip_code CDATA #IMPLIED           reporter_type  (single-multimer |
                         phone CDATA #REQUIRED                          multiple-oligomer |
                           fax CDATA #IMPLIED                           other)#REQUIRED
                         email CDATA #REQUIRED      other_reporter_type  CDATA #IMPLIED
                          href CDATA #IMPLIED >                 model  CDATA #IMPLIED
                                                               version  CDATA #IMPLIED
                                                                  href  CDATA #IMPLIED >
<!ELEMENT hardware         (contact?,
                            description?,
                            parameter*)    > <!ELEMENT seq_feature    (bio_seq?,
                                                                       ref_bio_seq?,
<!ATTLIST hardware         make CDATA #REQUIRED                        ref_clone?,
                          model CDATA #REQUIRED                        gene?,
                           year CDATA #IMPLIED                         reference?,
                           href CDATA #IMPLIED >                       treatment*,
                                                                       parameter*,
                                                                       coordinate?,
<!ELEMENT software         (contact?,                                 description?,
                            hardware*,                                 comment*)        >
                            description?,
                            parameter*)    > <!ATTLIST seq_feature    id ID   #REQUIRED
                                                                       name CDATA #IMPLIED >
<!ATTLIST software           id IDREF #REQUIRED
                           name CDATA #REQUIRED <!ELEMENT bio_seq       (reference |
                        version CDATA #IMPLIED                          CDATA)           >
                           year CDATA #IMPLIED <!ELEMENT ref_bio_seq   (reference |
               operating_system CDATA #REQUIRED                        CDATA)           >
                           href CDATA #IMPLIED > <!ELEMENT gene        (reference |
                                                                       CDATA)           >
                                               <!ELEMENT ref_clone     (reference |
                                                                       CDATA)           >
                                               <!ELEMENT coordinate    EMPTY            >
                                               <!ATTLIST coordinate  horizontal CDATA #IMPLIED
                                                                       vertical  CDATA #IMPLIED   >
```

Figure 5.10
The MAML pattern document type description. Shown are three fragments of the MAML DTD. The first column demonstrates the generic description of a data set creator. The second column demonstrates the capability of MAML to support composite expression measurements and analyses of the sort found on the Affymetrix platform.

In addition to the OMG efforts, there is the Interoperable Informatics Infrastructure (I3C), a group that is tasked to produce technical recommendations for the interchange of data with the goal of creating practicable solutions.[11] The membership of I3C is constituted of 60 companies and organizations committed to addressing the balkanization of data modeling efforts illustrated by table 5.2.

Again, it is interesting to note how much more difficult it has been to arrive at a consensus model for microarray expression measurements than it took to create a standard data model for genetic sequences in GenBank. The reason, as stated before, is that gene expression measurement is an analog measurement rather than the digital measurement of DNA sequence and therefore the measurements can only be understood with respect to a particular measurement system, a human artifact. The goal of the standardized microarray data model can only be realized once a common set of abstractions can be reliably found across all these different measurement technologies and artifacts.

5.5   Nomenclature

The issuing of consistent names to genes is an apparently mundane chore. Unfortunately, at this time, its importance may exceed that of shared data models and ontologies for the purposes of the immediate progress in our functional genomics efforts. Why might this be?

First, we want to make sure that we compare the behavior of the same genes and their properties across all experiments. When a microarray manufacturer reports that the nth element of its array interrogates the expression of a particular gene, the following questions should be asked: Which particular polymorphism(s) of that gene, which alternative splicing product(s), and which other gene(s) are being reported on?

Consequently, the nomenclature (*i.e.*, naming scheme) used to identify gene expression measurements on microarrays has to be able to capture the increasingly polymorphic definition of genes. As we learn more about single nucleotide polymorphisms (SNPs), alternate splicing products, pseudogenes, frame-shift mutations, and other variations in just the primary structure of the genes and the proteins they produce, a nomenclature has to be able to distinguish between these various forms of the same gene and yet group them as belonging to the same coding gene. This goes beyond the UniGene[12] effort which attempts to cluster known experi-

---

[11]http://www.i3c.org.
[12]http://www.ncbi.nlm.nih.gov/UniGene/.

mentally derived sequences into a cluster with a single consensus sequence, ideally corresponding to a single particular gene.

The task of finding a standardized and universal nomenclature for each gene is just the tip of the iceberg of a recurrent challenge in microarray technology: the task of linking a spot on a microarray to a specific gene or expressed sequence tag (EST). This is challenging in both mundane technical ways and in deep fundamental ways. The mundane ways in which it is challenging is that if one does not have the exact sequence for each microarray spot, at the very least, one has to obtain the correct sequence that each spot on the microarray was engineered to represent. For instance, on a spotted cDNA microarray, one has to determine the gene sequence of the clone used for each spot. With an oligonucleotide microarray, one has to determine from the manufacturer what sequence each probe set was designed to correspond to. In cDNA microarrays, these probes are typically linked to an IMAGE (see the description of the IMAGE consortium on p. 245) sequence ID. These are typically issued with the cDNA microarrays or with the probes that are spotted onto the microarray. Unfortunately, there has been a high error rate in many of the cDNA clones obtained from various manufacturers and so although the name mapping may be relatively straightforward, it may not actually represent what is being hybridized. Consequently, researchers may have to resequence a clone if they want to verify that the probe is assaying for the presence of the gene they think it is. The challenge is somewhat different with proprietary oligonucleotide microarray manufacturers, where the manufacturer's catalog states what gene each probe set has been engineered to hybridize to but the actual oligonucleotide sequences themselves are unknown.[13] This is problematic for several reasons.

- The gene sequence that the probe set may be engineered for may be a fusion gene, *i.e.*, an abnormal composite of two different genes. Consequently, since we do not know which part of the fusion gene is actually being represented by the probe set, it is unclear whether that probe set measures just the presence of the fusion gene transcript or whether it can also measure the presence of either one of the constituent genes separately.

- Furthermore, at the time of this writing, some oligonucleotides probe sets are designed against large sections of the genome such as a cosmid (see glossary, p. 277). Knowing the genomic sequence does not provide very much information about which gene in particular is being expressed from that region of the genome. If the

---

[13]That is, they are kept as confidential proprietary information.

specific oligonucleotides were made public, then one could perform a BLAST search to find which specific gene in that region the oligonucleotides corresponded to.

- As the genome efforts in the human, mouse, and other species increase in accuracy and completeness, the mapping between the microarray oligonucleotides and the genes they correspond to is likely to improve. Specifically, as we gain more knowledge of polymorphic variants of genes and their alternate splicing products, the investigator will be able to verify which particular variant a probe set is measuring. However, in the absence of such information, researchers cannot take advantage of increased knowledge of the genomic sequences to provide a more accurate labeling of the intensities that they are measuring on the microarrays.

- A related problem occurs with oligonucleotide microarrays where sequences that were thought to be part of one gene at one time are found to be components of other genes as the UniGene (see p. 244) clustering become more refined. Practically, this means that in one generation of Affymetrix microarrays, a probe set is said to correspond to one gene, whereas in another generation it may correspond to two or more genes based on the new assignments of the particular GenBank subsequences to different genes. In the absence of knowledge of the exact oligonucleotide sequences, or in the absence of a version history of Affymetrix accession numbers, the analysis of oligonucleotide microarray data, particularly across generations of microarrays, becomes problematic.

- Occasionally, manufacturers of oligonucletide sequences just make mistakes in their design process and engineer oligonucleotides that do not interrogate for the expression levels of the specified genes. The most recent and notorious example of such an error was in the design of Affymetrix oligonucleotide microarrays for murine systems that was unreported for months. If these sequences had been made public, then the error would likely have been discovered much earlier.

Pragmatically then, in the year 2002, some relatively laborious cross-indexing of terms across nomenclatures needs to be done in order to first determine which gene(s) a particular spot on a microarray was engineered to assay, and secondarily, to compare results across different microarrays. In practice, this means looking up in a table provided by each microarray manufacturer each accession number corresponding to each probe or probe set on a microarray, whether it be a tentative consensus (TC) number from the Institute for Genomic Research (TIGR) organization, a GenBank accession number, a UniGene accession number, an IMAGE clone ID, or some other accessioning scheme, finding the representative genetic sequence corresponding to that probe or probe set, and then running a Basic Local Alignment Search Tool (BLAST) search against one of the standardized nomenclatures such as LocusLink (see p. 243), if it is a gene with a known function, or UniGene, if a more comprehensive search is desired. Because this lookup function has to be done hundreds of times per analysis due to the large number of genes present, it requires one to write a script that submits the appropriate sequences to a BLAST program. The submission can be over the Internet using the batch version of BLAST at the NCBI. For better performance, if one has sufficient storage space locally, the entire GenBank repository and BLAST programs can be downloaded and run locally, using instructions at the NCBI. It is prudent to re-run these scripts on a regular basis as the UniGene and LocusLink databases are frequently updated.

An alternative approach is to develop the infrastructure to track each microarray probe or probe set and automatically determine the latest meanings for each. We have developed such a system that automatically parses the latest tables converting GenBank to UniGene, and UniGene to LocusLink, as well as the content files for these databases. Using these tables, one can then connect each of these references to the original publicly available accession numbers provided by the microarray manufacturers. The advantages of this approach are that:

- One can immediately find official names and symbols for probes and probe sets, as well as known synonyms.

- One can take the opposite route and search for probes and probe sets by gene meaning or other characteristics. For instance, if one has prior information that a gene on human chromosome 4 is associated with a particular phenotype, one can optimize functional genomics analysis by restricting to only those genes known to be in the appropriate location. One could also restrict analysis by protein domain, such as restricting to only those genes containing a zinc finger or DNA binding domain.

- One can find and search by functional categorization, such as a term from GO or a disease from Online Mendelian Inheritance in Man (OMIM).

- Because UniGene is essentially formed by applying BLAST to GenBank entries, it spares one from performing those BLAST operations.

- The tables can be updated regularly automatically.

We maintain a website at http://www.unchip.org which uses these tables and allows the translation of a microarray-specific accession number (currently mostly

Affymetrix) to one of the more globally used nomenclatures described below. A similar service is available for Affymetrix customers [14]. A commercial product offering similar, but more broadly applicable functionality is called GeneSpider and may be available soon from Silicon Genetics.

5.5.1   The unique gene identifier

Given the aforementioned desiderata that motivate the need for a standard nomenclature for genes, the quest for such a standardized nomenclature has generated widespread international efforts. There are several candidate nomenclatures, the most popular and useful of which we summarize here. Optimally, these nomenclatures should capture polymorphisms, splice variants, and mutations of the genes as they appear in one species and in their orthologs in all other sequenced genomes of other organisms. We list some of the more popular nomenclatures used for microarray experiments. In practice, microarray repositories will use several nomenclatures to reference the measurements made in gene expression experiments. For a more comprehensive and current compendium of nomenclatures such as these, the reader is referred to issue no. 1 of *Nucleic Acids Research*, of any recent year.

LocusLink (`http://www.ncbi.nlm.nih.gov/locuslink/`)   Perhaps the most extensively curated and possibly the broadest nomenclature is maintained through the LocusLink database [145]. LocusLink identifiers are stable, in that they are designed not to change over time, and are genome independent. There are other databases which have a much larger set of expressed sequence tags (see glossary, p. 277) or putative genes, but for which either the complete coding region is not known, or for which the function of the gene product is unknown. The curated nature of LocusLink is important because, unlike GenBank records, which are under the full editorial control of only those who submitted the sequence, the sequences and annotations in LocusLink are under an extended and distributed editorial process.

Broadly speaking, the LocusLink record is comprised of three components. The first is a stable unique identifier called the locus ID. The second component is the reference sequence which is obtained from the REFSEQ[15] database, which in turn is seeded by a GenBank source coding sequence. This reference sequence is left as provisional until the full length is obtained according to the editorial process. The third component includes all the annotations such as the completeness indicator for the gene (indicating that the entire coding component of the gene is complete),

---

[14]`http://www.netaffx.com`
[15]`http://www.ncbi.nlm.nih.gov/LocusLink/refseq.html`.

accession numbers providing links to a protein record, links to the OMIM,[16] links to other databases such as UniGene and DBSNP, and an extremely useful English-language summary that describes the gene. Of note, LocusLink is involved in an active collaboration with the Human Gene Nomenclature Committee (HGNC)[17] to provide an overall unifying international standard. Most recently, LocusLink also maintains links to functional annotation databases such as those supplied by Proteome[18] and publicly maintained ontologies such as GO.

It appears likely that LocusLink will continue to be the locus for much annotation of organismal genomes and the provider of stable identifiers for their genes.

UniGene (http://www.ncbi.nlm.nih.gov/UniGene/)   Although LocusLink provides the most stable identifier for known genes, or at least those for those which at least one function is at least presumed, at this early stage of the discipline of functional genomics, the majority of genes do not have any known function.[19] For this reason, other related databases are useful. Chief among these may be the Uni-Gene system at the NCBI [28]. UniGene attempts to be comprehensive; to catalog all the characterized genes as well as hundreds of thousands of uncharacterized and novel expressed sequence tags, while attempting to reduce the redundancy inherent in those individual catalogs. The content of each UniGene record includes a set of sequences with GenBank identifiers that have been clustered such that they are presumed to correspond to a single specific gene. As this clustering is done mostly automatically, and the clustering procedures have evolved over time and been augmented with more data, the UniGene clusters are not guaranteed to be stable, unlike LocusLink identifiers. The strength of UniGene is that it attempts to be comprehensive and includes not only the GenBank accession numbers for sequences in each cluster but also known alternate splicing variants. Currently, UniGene describes human, rat, cow, mouse, and zebra fish organisms. Each Uni-Gene record has a unique ID called a cluster ID. Because these cluster identifiers are unstable, a cluster ID can be retired for a variety of reasons.

- The sequences that congregate to a cluster might be found to be contaminated.

- Two or more clusters may be joined to form a single cluster in which case the original cluster IDs are retired.

---

[16]http://www.ncbi.nlm.nih.gov/Omim.

[17]http://www.gene.ucl.ac.uk/nomenclature.

[18]http://www.proteome.com.

[19]And a larger majority only have a few of their functions known.

- A cluster may be found to be composed of more than one gene and therefore must be split into two or more clusters. Thus the original cluster ID has to be retired and new ones generated for the subclusters.

Because of this terminological instability most reports using the UniGene nomenclature will also report the "build" (*i.e.*, version) number from which the cluster IDs were drawn.

GenBank (`http://www.ncbi.nlm.nih.gov/`) An overview such as this would be incomplete without mentioning GenBank, the public database of all known nucleotide and protein sequences from which many of the other nomenclatures are derived [22]. At the time of this writing, GenBank contains over 12 million annotated sequences and is growing at the rate of over 5 million sequences per year. GenBank sequences are divided into divisions, roughly corresponding to taxonomy (*e.g.*, bacteria, viruses, *etc.*) as well as for specific projects, such as ESTs, genome sequencing, sequence-tagged sites, and others.

Each GenBank record contains taxonomic information regarding the organism from which the sequence was obtained, bibliographical references, and biological features found within the sequence. Each record has an accession number that is stable and unique; each entry is also assigned a unique unstable GI number so that revisions may be tracked. Beyond this, very little additional annotation is provided, given the low-commitment nature of this database.

GenBank is a member of the International Nucleotide Sequence Database Collaboration, including the DNA Data Bank of Japan and the European Molecular Biology Laboratory, which defines common ontologies for taxonomy and features allowing daily movement and translation of data between these databases.

IMAGE (`http://image.llnl.gov/`) The Integrated Molecular Analysis of Genomes and their Expression (IMAGE) consortium was formed to create, collect, and characterize cDNA libraries from a variety of different tissues and was initially spearheaded by the efforts by the Washington University Genome Sequencing Center and Merck & Co. [118]. Several microarray databases refer directly to the IMAGE clone identifiers rather than to any one of the more precisely characterized sequences described above. Those IMAGE cDNA clones that have been sequenced are available in the dbEST database maintained at the NCBI.[20] Perhaps the best feature of the IMAGE clone identifiers is that they refer to physical instances of these clones on the master plates from which the clones are generated. In that sense, the

---

[20]`http://www.ncbi.nlm.nih.gov/dbEST`.

clone identifiers are stable. However, considering that some of the clones obtained
from these master plates have been known to suffer contamination or have not been
sequence-validated, they are not stable.

The Institute for Genomic Research (TIGR, http://www.tigr.org/)   TIGR
also maintains a fairly extensive list of human and other organism nomenclatures.
TIGR has its own consensus curation and assembly process that it uses to maintain
a number of databases for organisms, including human, mouse, rat, zebra fish, rice,
Arabidopsis, soy, tomato, maize, potato, cattle, and many others. The nomenclature specific to TIGR is the tentative consensus sequence, or TC. TC clusters are
created by a process similar to UniGene by assembling ESTs into clusters based
on sequence. However, alternate splice forms are built into separate TC clusters,
as opposed to UniGene, where these are kept in the same cluster. Many manufacturers of microarrays refer to TIGR accession numbers or TC identifiers in their
descriptions of the probe sets or cDNA probes on their microarrays. TIGR itself
uses TC identifiers to create a useful set of mappings of orthologs across all the
species indexed in the TIGR database.

Enzyme nomenclature database   The Nomenclature Committee of the International Union of Biochemistry and Molecular Biology maintains a catalog of known
enzymes and their functions. Enzyme nomenclature is based on the reactions that
are catalyzed, and not the genes that make up the enzymes, or the protein structures of those enzymes. Thus, simple one-to-one translations between Enzyme
Commission (EC) numbers and GenBank or LocusLink accession numbers are neither possible nor accurate. For example, the enzyme described by EC 1.1.1.27
(lactate dehydrogenase, LDH) assists in the conversion of lactic acid to pyruvate.
However, that enzyme is made of four independently chosen subunits of two types,
either LDHA (LocusLink 3939) or LDHB (LocusLink 3945). Note that these genes
have different LocusLink entries in other species, whereas the enzyme function keeps
the same EC accession number.

   This nomenclature can be searched using the Expert Protein Analysis System
maintained by the Swiss Institute of Bioinformatics [21].

Other identifiers   Other identifiers used on microarrays include the Expressed
Gene Anatomy Database identifiers [22], and the Munich Information Center for
Protein Sequences Yeast Genome Database [23]. In addition, other identifiers of

---

[21]http://ca.expasy.org/enzyme/
[22]http://www.tigr.org/tdb/egad/egad.shtml
[23]http://www.mips.biochem.mpg.de

which a functional genomicist should be aware include Ensemble [24], the Genome Database [25], the Mouse Genome Database [26], the Saccharomyces Genome Database [27], and the Database of Transcribed Sequences [28].

5.6   Postanalysis Challenges

5.6.1   Linking to downstream biological validation

When we introduced the functional genomics pipeline of figure 1.5 (see p. 17) we made it clear that without biological validation, the hypothesis generation and test procedure for which the pipeline was intended will be arrested. A small but important subtask of bridging the flow from computational to biological validation is the assurance that the genes referred to in the analysis (*e.g.*, the lists of genes belonging to each cluster) should refer to the appropriate physical clone regardless of whether the nomenclature used for analysis was GenBank, UniGene, TIGR, Stanford Yeast ORL codes, or any other internal or local numbering schemes. Specifically, biological validation will involve performing biological "wet bench" experiments, which may well depend on obtaining clones of the genes identified in the analysis.

Thus, some degree of translation is necessary to follow up on the results. It has been our experience that this process takes manyfold longer than the actual bioinformatics analysis. Since many of the genes or ESTs that appear on the lists after analysis are typically unknown, one can argue that a bioinformatics analysis of microarrays is never actually complete. It may only be after another week, month, or year, that another group may finally assign a meaning to one of those unknown genes. Only after the analyst sees that result might a hypothesis be generated. Thus, an infrastructure that periodically re-queries the accession numbers or genes to see if anything new is known about each oligonucleotide, cDNA, or gene is necessary. Furthermore, appropriate analysts should be informed when such an event happens. Such a system might be similar to how PubCrawler operates on the biomedical literature [96].

Here we touch upon some of the conceptually shallow but pragmatically thorny challenges that face the researchers as they attempt to make sense of the results of their data mining forays.

---

[24] http://www.ensembl.org
[25] http://www.gdb.org
[26] http://www.informatics.jax.org/mgihome
[27] http://genome-www.stanford.edu/Saccharomyces
[28] http://www.allgenes.org

5.6.2   Problems in determining the results

Several problems can develop during the postanalysis interpretation of these results:

- No genes BLAST to finding: Some microarray results may be given as accession numbers for ESTs. Thus, to follow up these results, it is necessary to take the nucleotide sequence for the EST and BLAST that sequence against GenBank or UniGene looking for a "full-length" gene. In translating these accession numbers, however, there may be no gene that matches strongly enough. This can happen because the full-length gene may not be known yet, or due to inaccuracies in the EST consensus sequence. Gene findings with this problem may not be able to be followed up.

- Many genes BLAST to finding: Sometimes, many full-length genes may be found for a single EST, meaning the EST is nonspecific. At the time of the microarray construction, the EST may have been thought to be specific for one gene, but with progress in sequencing, this may no longer be true. To follow up these findings, many clones may need to be used in biological experiments.

- EST is no longer unique: Often, microarrays may have been designed with a particular sequence thought to be unique for a single gene, EST, or EST cluster. With progress in sequencing, however, it may be later realized that what was thought to be a single EST or gene cluster actually represents two or more unique clusters. It is often unclear as to which of the new clusters best maps to the older set of probes. Unfortunately, revision control is often not comprehensive in these accession number databases, and it becomes very difficult to link new clusters with old designators.

- Genomic sequence BLASTs, but not genes: Sometimes, gene sequences from positive microarray findings can be found in stretches of genomic sequence using BLAST. Additional bioinformatics (and biological) work then needs to be done to determine the gene boundaries before biological validation can take place.

6   From Functional Genomics to Clinical Relevance: Getting the Phenotype Right

Functional genomics experiments can only have biological or clinical relevance if information (or better yet, knowledge) external to the system being studied is employed. Even in the case of so-called unsupervised algorithms or clustering techniques, the early investigations (*e.g.*, [56, 63]) made sense of the clusters of expression only by reference to external knowledge: the genes known to be involved in cell cycle regulation, protein translation, *etc.* Without such external *a priori* knowledge, the observed clusters are of little use in generating or testing new hypotheses about biology. In the experiments that had the most obvious clinical relevance, such as finding genes that aid in the distinction between acute myelogenous leukemia versus acute lymphoblastic leukemia [78], the prognosis of large B-cell lymphoma [3], or pharmacogenomic prediction of gene targets for chemotherapy [39], it was central to the import of these investigations that they integrated well-characterized external biological or clinical data with the expression data.

If functional genomics is to lead to a qualitative change in the way in which clinical medicine (diagnostics, prognostics, and therapeutics) is practiced, then most studies will similarly have to incorporate high-quality and relevant external data. That is, if we want to know how a gene expression pattern may predict mortality, a particular disease profile, or drug responsiveness, we need to be at least as meticulous in characterizing the patients, their histories, their tissues, and the contexts surrounding how data were acquired, as we are in obtaining the gene expression profile. Perhaps because the need has only become recently apparent, to date there has been little in the way of systematic approaches to the acquisition of such extragenomic data. We discuss here why this is and what steps can be taken toward a more systematic approach.

6.1   Electronic Medical Records

A central intellectual and technological asset to functional genomics has been GenBank (see section 5.5.1) and related genomic and protein databases. Their standardized data models have allowed research laboratories throughout the world to rapidly populate them with the very latest information. In turn, these databases are freely available throughout the world via the Internet and have seeded, accelerated, and inspired thousands of research projects.

In contrast, there are few, if any, consequential shared national clinical databases. Specifically, patient data in one information system can only rarely be transferred to another system in clinical practice. Within a single healthcare system, annotations such as billing diagnoses will suffer from distortions imposed in an attempt to maximize reimbursement for care. Despite decades of research and development of clinical record systems, they remain problematic. If properly constructed and maintained, electronic medical records could provide an invaluable set of phenotypic annotations with which to bring clinical relevance to genomic data.

The marked contrast between genomic and clinical care databases is deceptive. The Human Genome Project has benefited from the elegant simplicity of the genetic code. Consequently, there are relatively few items that GenBank requires to be submitted for a entry to be a valid (and useful) component to its database. The clinical care of human beings is a far more complex process, requiring at the minimum a detailed record of the history of multiple clinical interventions and outcomes, relevant life history, and clinical measurements that span several modalities, from serum chemistry to brain imaging. It is not surprising that the data model required to capture all this information is extremely complex as is evidenced by the Health Level Seven (HL7) Reference Information Model [147]. It is a remarkable tribute to the persistence of the individuals involved in these clinical data model standardization efforts that they have been able to arrive at a reasonably adequate standardized representation of not only the aforementioned descriptors but much of the process and business of clinical care.

This challenge relates back to our prior discussion of the complexity of microarray expression data[1] which is much more akin to that of clinical laboratories than that of DNA sequences. Microarray measurements reflect the particular artifactual qualities of the measurement system used, in contrast to the (at least theoretical) invariance of sequence with sequencing technique. It is therefore not surprising that even prior to encompassing the entirety of clinical annotation, the genomics community has faltered in developing shared and standardized data models where the simplicity of the genome no longer dominates. However, the complexity of microarray data models is easily dwarfed by the complexity of clinical data, and the effort to generate a standard human phenotypic data model for genomics would similarly demand orders of magnitude more effort than has been invested in data modeling by functional genomicists to date. It is therefore somewhat surprising that several efforts to develop clinical phenotype data models have been proposed without much reference or attempt to co-opt the efforts of the clinical informatics community.

A more efficient mechanism might be for functional genomicists and bioinformaticians to appropriate that segment of an existing clinical data model, such as HL7.

---
[1] As we described in section 5.4.

This might require that members of the bioinformatics community participate in an HL7 committee in order to fully accommodate the annotation needs of functional genomics. This would be a far smaller cost than having to duplicate the existing massive efforts of HL7.

6.2 Standardized Vocabularies for Clinical Phenotypes

Standardized clinical vocabularies are just as essential as standardized phenotypic data models. The lack of widely accepted standardized vocabularies for clinical care has greatly hampered the development of automated decision support tools and clinical research databases. The impossibility of guaranteeing that a serum sodium or systolic blood pressure has the same code or term throughout our hospital system is troublesome. Fortunately, several efforts in the private and public realm, *e.g.*, Logical Observation Identifiers, Names, and Codes (LOINC) [99], are addressing this issue. The National Library of Medicine has invested large resources in the Unified Medical Language System (UMLS) to enable these different vocabularies to be interoperable, at least at a basic level [120, 158]. Most of the most widely used vocabularies, such as Reed codes [84], Medical Subject Headings (MESH) [51], and Systematized Nomenclature of Medicine (SNOMED) [53] are represented in the UMLS.

The same problems are not unknown in bioinformatics as is pointed out in the section on nomenclature (section 5.5). Many microarray technologies use their own system of accession numbers, which then have to be translated into a more widely used nomenclature such as LocusLink.

As is the case for data models, as functional genomic investigations venture into the domain of clinical correlates of gene sequence, expression, and protein concentrations, the lack of standardized clinical vocabularies will prevent large-scale accumulation of useful data at the national or even regional level. For example, if a set of expressed genes are thought to be associated with a particular set of cardiac arrhythmias, this association will only be found if the nomenclature used to describe the arrhythmias in the clinical databases is drawn from one of the standardized vocabularies. Additionally, these vocabularies would have to be used with the same semantics at each data collection site. Rather than inventing yet another competing standard for a clinical vocabulary, bioinformaticians should examine which of the existing component vocabularies within the UMLS best suits the needs of a particular set of studies.

6.3   Privacy of Clinical Data

In 1997, the Institute of Medicine [47] reported on significant lacunae in both technology and policy in protecting confidential patient data. Among the problems
of most concern emphasized in this report was the relatively unrestricted access
by third parties to these data for secondary uses and the lack of the adequacy of
the anonymization process (in practice and theory [170, 171]). Subsequent to this
effort, the clinical informatics community developed several model confidentiality
policies [153] and cryptographic identification systems [111, 174].

As the fruits of the Human Genome Project are translated first into clinical research protocols and then into clinical practice, personally identifiable genomic data
will find their way into several instances of the information system. The challenges
posed to security and privacy of such data might seem to dwarf any encountered to
date with conventional clinical data. At first blush, genomic information is likely
to be much more predictive of current and future health status than most clinical
measurements. And, with very few exceptions, an individual's genome is uniquely
identifying and applicable to that individual's entire life. This identifiability is
apparently more reliable, persistent, and specific than typically cited identifiers,
including a person's name, social security number, date of birth, or address. In
practice, however, the widespread nature of clinical databases and the ease with
which a unique individual can be identified [170] allow for breaches of privacy to
be effected relatively trivially, even without genetic information. Furthermore, as
demonstrated by insurance companies, highly accurate actuarial and prognostic
knowledge can be developed solely on the basis of clinical history, habits such as
smoking, family history, a physical examination, and a handful of laboratory tests.
It remains to demonstrated, given all the genomic reductionist fallacies reviewed at
the beginning of this book, just how much additional prognostic accuracy will be
delivered by the use of personal genomic data.

Nonetheless, at the very least, the architects of information systems storing genetic data should learn from all the mistakes and designs developed for the security
architectures and privacy policies of conventional clinical information systems. Conversely, the dire concerns voiced over the storage of personal genetic data will likely
generate new policies and security architectures that will enhance the confidentiality of clinical information systems. Moreover, when personal genetic data become
incorporated into routine medical practice, the safeguards in place for the medical
record's confidentiality will have significant import for the confidentiality of the
genetic data that will be referenced in the medical record. It also may further drive
the increased degree of patients' control of their own medical record [125].

Why should we bring up these concerns at all in a book on functional genomics? Simply because there are two strong imperatives to keep the privacy of the patient or tissue donor intact. First, on the simple grounds of human rights, there is a basic or implicit contract between researchers and patients. The second and less altruistic motivation is that if the confidentiality of the patient is breached, even the perception of the loss of confidentiality (particularly if it involves a disease for which there is attached social stigma or risk of loss of employment or income) could jeopardize an entire study. Or it could well lead to a broad societal reaction against genomic studies. Even now, there exist multiple recourses for the patient to obtain both civil and criminal penalties against the researcher. These penalties have further been formalized by the recent components of the Health Insurance Portability and Accountability Act (HIPAA) as amended in the final days of the Clinton administration in 2000 [32].[2] In this brief section we outline the major obligations involved in protecting the confidentiality of genomic patient data and common pitfalls in implementing them.

6.3.1 Anonymization

In theory, if a database were to be truly anonymous, then, even if it fell into the possession of an unscrupulous party, no harm could come of it. Therefore, a first attempt at the design of a functional genomics database will typically involve one of the following two strategies for anonymizing a database.

The first strategy is to delete or replace common identifying attributes, such as name, date of birth, or address, which could be used very easily with external databases to identify the patient.[3]. The risk with this approach is that although it does indeed reduce the ease of identification to eliminate these common "face-recognizable" identifiers, the records can still be readily identifiable. The underlying insight is that there will be a list of attributes, such as which doctor the patient visited, when the patient was first diagnosed, or what kind of cancer the patient had, which can serve to uniquely identify the patient. At the very least, it can be used to reduce the set of patients matching these attributes to a handful of individuals. Then, if there is an incentive to do so, there is little or no barrier to reidentification of the patient record. This approach has been most extensively developed and documented by Latanya Sweeney [169, 170], who documented that even meticulously deidentified records such as those from the billing systems of Medicare could be used to re-identify patients, including prominent politicians. She developed a col-

---

[2]And as pursued through at least the early months of the George W. Bush administration.

[3]A list of such identifiers that have been deemed of particular risk by the U.S. Department of Health and Human Services as part of HIPAA can be obtained at http://aspe.hhs.gov.

lection of programs, called SCRUB, which enable an increasing degree of anonymity due to elimination of data fields which could be used to particularly identify a single patient. Nonetheless, one of the clearer messages of that research was that to the degree that a clinical database is useful in providing clinical detail to characterize the patient, there is an increasing risk of breach of the anonymity of that record. With this in consideration, designers or architects of functional genomic databases should be mindful of being parsimonious in their choice of attributes used to annotate these clinical databases, such that only the minimum needed for the purposes for which the databases are designed should be implemented.

The second, alternative approach is to segregate those fields, including name and addresses, that would be the most revealing of the patient's identity and place them into a separate identification database controlled by a third trusted party (trusted by the patient) who would only allow the patient data to be "joined" to the nonidentifying data with explicit permission of the patient or the institutional review board of a research institution. That is, the reidentification of the patient would be determined by consent and disclosure policies and practice. We note that the anonymity of the nonidentifying data could still be compromised by the same reidentification techniques applied to the scrubbed databases described above.

In our own investigations of multicenter large-scale genomic databases, we have investigated the use of cryptographic health identification systems which could be used locally or nationally to create systems that enabled either the patient, the provider, or the institution to determine very specifically under which circumstances the patient record could be joined to the identifying information [111]. Furthermore, we have argued for patient-controlled encryption of their record [125] to prevent the application of reidentification techniques.

In the end, however, it should be recognized that no matter how secure the means used to encrypt the database or to anonymize it, if there is a sufficiently large financial or personal incentive to breach the privacy of a particular individual, it will be done, whether through a direct assault on the protection measures used, or through a corruption of internal controls be they technical or sociological. In this light, it is most reasonable to ensure that when patient samples are entered into a database, they be obtained only upon very extensive debriefing of the patient of the measures that will be used to protect the record, as well as the potential risks. With full disclosure and consent, there is much less likely to be acrimony and *post hoc* disruptions that may affect the integrity of the entire functional genomic investigation undertaken by a particular investigator.

6.3.2 Privacy rules

By passing the HIPAA, Congress established standards governing the privacy of identifiable health information. Though this legislation covers many aspects of privacy in clinical medical care, there is a significant area of overlap with clinical research and the practice of obtaining consent from human beings involved in such research.

Those organizations covered by the HIPAA legislation and the privacy rule can continue to use properly deidentified health information for research purposes. However, those organizations can only use and disclose *protected health information* for research purposes with permission from each individual, or without individual authorization in a restricted set of cases.

1. An institutional review board has approved a waiver of research participant's authorization. This could be done if all of the following conditions are met:

    - Use of the information involves no more than minimal risk to the individuals;
    - The waiver will not adversely affect the privacy rights of the individuals;
    - The research could not be done without a waiver or without use of the information;
    - The risk to the research subjects is low compared to the potential benefits to those subjects;
    - A plan is in place to protect against improper use;
    - Identifiers to the information will be destroyed as early as possible;
    - The information will not be re-used.

2. The researcher documents that the data will be used only to design or assess the feasibility of a research project.

3. The researcher documents that the protected health information of deceased individuals is being used and is necessary for research on their decedents.

This section is not meant to provide specific legal advice on this subject. More information about the implications of HIPAA on clinical research is available at http://aspe.dhhs.gov. There are other U. S. laws to protect research subjects, including the Common Rule from the Federal Drug Administration. Privacy rules in other countries, especially in the European Union, will differ. The reader is strongly encouraged to consult his or her own institutional review board, or equivalent organization, for further guidance on this subject.

6.4   Costs of Clinical Data Acquisition

"Getting the data in" has often been cited [130] by authorities in clinical informatics as being among the most difficult challenges in successfully deploying clinical information systems. In particular, the costs of acquiring detailed and structured data from the clinical care process have been daunting. Voice and handwriting recognition information systems have not been broadly adopted for a variety of performance and usability issues. These cost and practicality issues will continue to present an obstacle to clinical information system utility and deployment until better solutions are arrived at. In contrast, the Human Genome Project has managed to achieve significant economies of scale in data acquisition in sequencing and expression measurement. Gene microarrays alone have dropped in cost by a factor of 10 in just the last 2 years.

Here again, once genomic investigators attempt to bridge the gulf from purely genomic data sets to phenotypically (*i.e.*, clinically) annotated data sets, they will be confronted with the same challenges of clinically oriented, codified data acquisition. The questions of which user interfaces are the most cost-efficient, reliable, and generalizable to multiple clinical domains are among the implementation and design challenges that they will face. Although they have yet to arrive at definitively successful answers, clinical informaticians have already completed several decades worth of engineering and ethnographic studies [49, 50] addressing the very same questions.

We have personally witnessed the difficulties of some genomic studies in which the genotyping portion of the study was readily completed but the phenotyping was costly, slow, and ultimately rife with errors. This experience is likely to be increasingly common and therefore the need to solve the clinical annotation challenge is pressing.

7 The Near Future

In this chapter, we take the risky road of futurology with respect to functional genomics. We will make it even riskier by taking a short-term view, so that we will soon know just how far off the mark we were. The reason that we are at all engaging in this exercise is that the technological, biological, and analytic aspects of functional genomics investigations are changing every year. This leads to the, possibly vain, conceit that anticipation of future trends will allow a laboratory to invest in the most high-yielding technologies and personnel. We find ourselves asking the same questions many of our collaborators have asked us in this regard:

- What technology should we invest in for gene profiling measurement?

- What software platforms should we license, install, and become proficient in using?

- Even if we consistently use the same platform, will we be able to compare results of expression profiling experiments across different generations of the same microarray technologies?

- What are the most important issues we should plan for in the next 10 years to best leverage our efforts in functional genomics?

We attempt here to begin to answer these questions using the data on hand, our intuition, and our experience with past technological revolutions. *Caveat lector*, prognostication is all too cheap.

7.1   New Methods for Gene Expression Profiling

In the last 5 years of the twentieth century, and subsequently, there has been rapid growth in the development of alternative techniques for measuring gene expression on a large scale. There are two principal drivers behind the growth of diverse measurement techniques. The first is simply the commendable attempt to provide technologies that are more reliable, reproducible, precise, and cost-effective. The goal, after all, is to be able to make the measurement of a set of gene profiles no more expensive than a simple serum chemistry drawn routinely during a doctor's visit. It should be robust and affordable enough to be deployable widely for purposes of microbial genome detection for public health surveys and for routine screenings for a variety of diseases during clinical care and well-care checkups. Unfortunately, the world of commercial intrigue and murkiness around intellectual property has been the second important driving force in developing these technologies. Affymetrix, for

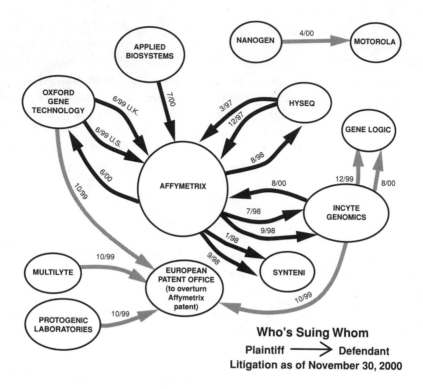

Figure 7.1
Lawsuits in the microarray field. With the number of lawsuits already in this area, it should be no surprise that companies in this industry are striving to determine new and unique intellectual property. (From Gibbs *et al* [77].)

example, has a patent that, under some interpretations, might include a range of microarray gene expression measurement technologies on a two-dimensional surface. Consequently, much of the development of alternative technologies has been focused on avoiding the potential conflict with this intellectual property (see figure 7.1).

We review here some of the major and upcoming technologies. Perhaps the best way to think of these technologies is within the following framework. Some of them use full-length cDNA probes, others use oligonucleotide probes. Independently of what kind of probe they use, the technologies may use a solid two-dimensional surface, a three-dimensional matrix, or optically coded beads.

We are hardly going to be able to even enumerate all the emerging technologies for expression measurement. For the latest up-to-date list, the reader is directed to http://www.gene-chips.com/. Also, for an excellent review of the scientific underpinnings of many of these technologies, we refer the reader to a review by Southern *et al.* [166] which provides a survey of the physicochemical basis of molecular interactions of hybridization on microarray surfaces. We would also impress upon the reader that most of the technologies described either have only begun early production lines or are still under development. Consider that it has taken well over 3 years, and counting, for the mainstays of microarray expression measurement, oligonucleotide microarrays, and robotically spotted microarrays to achieve minimal levels of accuracy and reproducibility. Then, as interesting and appropriate as some of the following methodologies may appear, they bear close scrutiny and testing prior to any large scale attempts for use in scientific experimentation.

Additionally, each of the technologies described is, of necessity, undergoing iterative processes of improved quality, higher density, and lower cost. This makes it all the more challenging to decide which microarray platform to adopt. This is very much analogous to choosing a home video gaming system: Nintendo, Sony, Microsoft, each has its own game console. As each manufacturer rolls out its next-generation platform at different times, consumers have to agonize over whether it is better to buy the platform that best suits their needs today or the one that will do twice as much in 4 months. As the refinement of these platforms is an iterative, continual, and competitive process, there is no stable or consistent "correct" response to the consumers' dilemma.

The video gaming console analogy is also relevant to the commoditization of microarray platforms. In the 1980s, real-time three-dimensional rendering was the province of highly specialized, expensive computer systems available in only a few laboratories. The cost of the hardware and staff far exceeded the cost of the system software. Now, with commodified video consoles, over the short life of a consumer system, the major cost is the acquisition of software and game modules. Similarly, the current microarray systems are labor-intensive and the hardware (including the microarrays themselves) and staff drive the high costs of expression profiling laboratories. As a result of the competition and improvements in manufacturing technology, the costs of hardware and attendant staff will diminish and likely become dominated by the costs of tissue acquisition, banking, and annotation.

With these issues in mind, we review some of the available and upcoming platforms.

7.1.1   Electronic positioning of molecules: Nanogen

Most biological molecules have a natural positive or negative charge. When these molecules are exposed to an electric field, those with a positive charge move to an area with a negative charge, and *vice versa*. Novel microarray technologies such as those of Nanogen (San Diego, CA) use this property to maneuver molecules to specific test sites on an electronic microchip. By carefully controlling which element of a grid of electrodes is charged, this technology can direct many test molecules to specific areas of the chip. Because of the more active nature of this technique, this could potentially deliver molecular binding on the microarray several orders of magnitude faster than passive hybridization methods.

Regarding gene expression measurement, a cDNA molecule that is negatively charged moves in an electric field to an area of net positive charge. The sample DNA is significantly concentrated over time in the area of positive charge. DNA that does not have the right mass/charge ratio is repelled from the area of the electrode under closely controlled electronic conditions (see figure 7.2). The current trade-off (pun intended) for all this configurability is that the density of features of these microarrays is at least one order of magnitude less than the current density offered by companies such as Affymetrix. However, the density of this technology is rapidly growing.

This system may be best suited for laboratories that are interested in a medium-throughput platform that will allow several genomic applications such as gene profiling, proteomics, and genotyping.

7.1.2   Ink-jet spotting of arrays: Agilent

Agilent (Palo Alto, CA) is building on its long history of highly engineered ink-jet delivery systems. It has licensed technology from Rosetta Inpharmatics (Kirkland, WA) so that it can synthesize and "spray" onto a glass slide oligonucleotide molecules in a highly localized fashion (see figure 7.3). That is, the nucleotides are "sprayed" separately by the ink-jet heads and linked *in situ* on the glass slide into oligonucleotides. At present, this technology permits up to 25,000 features per glass slide using volumes only in the picoliter range per feature. Because the features are built using longer nucleotides, there is typically only one feature per gene. This is in contrast to the 40 different features that are used in each probe set on Affymetrix oligonucleotide microarrays, using shorter and therefore individually less specific oligonucleotides.

In theory, the contact-free nature of the ink-jet process could result in fewer spatial artifacts than the robotically spotted technique. Also, the synthesis technique used allows oligomers with a length up to 60 nucleotides to be constructed. Again, in theory, this could provide more specificity than the shorter oligonucleotides used by Affymetrix, and better normalization by GC content and correction for cross-hybridization than longer cDNA molecules. The nature of this technology allows for a very rapid design and test cycle for these microarrays. Although it remains

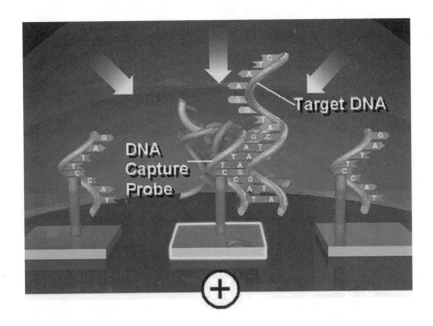

Figure 7.2
The Nanogen microelectronic array.

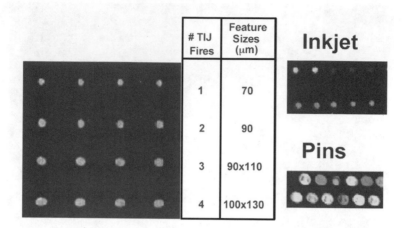

Figure 7.3
Flexibility of on-the-fly ink-jet arrays. On the right is a comparison of morphology of ink-jet versus pin arrays. Ink-jet printing produces uniform and consistent features. An artifact of on-the-fly printing is formation of oblong spots at maximum print speed (full fires) that can be made into circles upon slowing down. This underscores the theoretical flexibility of this system.

to be seen (at the time of this writing) whether Agilent does deliver on the promise of this technology, it has an opportunity to create a high-quality platform with manufacturing costs orders of magnitude less than others.

7.1.3 Coded microbeads bound to oligonucleotides: Illumina

Technology from Illumina (San Diego, CA) uses oligonucleotides linked to beads which are individually optically coded. These beads are then bound to the tips of individual fibers in a fiber-optic bundle (see figure 7.4). The company states that currently 2000 different genes can be represented on each fiber bundle, but appears to be scalable from 50,000 to 5 million beads per bundle. In addition, each array contains between 96 and 1536 bundles. Each bead can be optically interrogated by laser excitation via the fiber-optic bundle, using a proprietary technique. This technology is very much in the early testing stage and we will not have a good idea of its reproducibility and accuracy until after more independent testing.

If this technology does in fact prove viable, then its potential scalability to millions of probes will allow new ground to be reached in profiling. For example, millions of probes would allow each gene in the human genome to be represented, as well as all known polymorphisms and alternative splicing products.

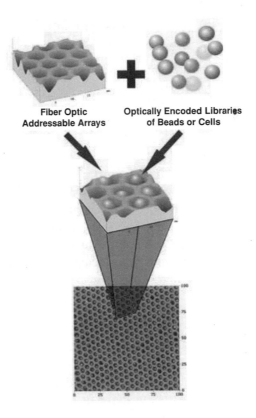

Figure 7.4
The Illumina BeadArray.

7.1.4   Serial Analysis of Gene Expression (SAGE)

SAGE is a technique designed to take advantage of high-throughput sequencing technology to obtain a quantitative profile of cellular gene expression [183]. Rather than directly measuring the expression level, the technique "counts" the number of transcripts or "tags" for each gene. A tag is a specific oligonucleotide of length 9 to 11 nucleotides corresponding to the transcript of interest which is selected on the 3' end of a transcript, relative to a site recognized by a restriction enzyme (usually NlaIII). First, double stranded cDNA is made from mRNA, using a biotinylated oligo(dT) primer. The cDNA is then cut with the anchoring enzyme (NlaIII), expected to cut every at every 256 base pairs. The resultant double-stranded cDNA is bound to streptavidin beads, divided into two pools, and ligated to one of two linkers. The tags are then spliced out using a tagging enzyme that cuts at a location away from its recognition site (such as FokI). The tags are joined together and the resultant string of tags is sequenced. This sequencing allows each SAGE tag to be counted. Therefore, the number of tags for each gene that the tag represents should be proportional to the expression level of that gene. Figure 7.5 illustrates this process.

SAGE suffers from the potential lack of specificity of these oligonucleotide tags, much in the same way as do the oligonucleotide and cDNA microarrays. It can be argued that the small length of the tags makes them even more susceptible to non-specific interactions (*i.e.*, multiple genes or gene products with the same sequence as the oligonucleotide). In fairness, the most recent gene profiling investigations use longer nucleotide sequences. Furthermore, and more problematically, sequencing technology has an error rate that might substantially degrade the correct assignment of a tag count to a gene. Although there have been some early, optimistic results for small numbers of genes, showing the correlation of SAGE results with microarray expression values [100], more comprehensive testing is necessary. Of course, as with all these technologies, the choice of a "gold standard" to rate such comparisons is hard to come by.

7.1.5   Parallel signature sequencing on microbead arrays: Lynx

Lynx Therapeutics (Hayward, CA) combines two techniques to assess gene expression: the ability to sequence without using a gel, and the ability to clone millions of templates onto microbeads [31]. Specifically, cDNA templates are first ligated into vectors containing a set of different 32-mer oligonucleotide tags. A 1% subset of these ligated vectors is taken, such that each template is ligated to a unique tag and each template is represented at least once. The subset is then PCR-amplified,

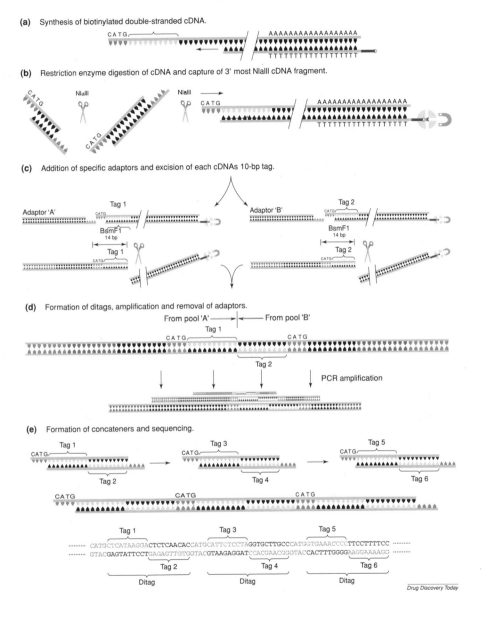

(a) Synthesis of biotinylated double-stranded cDNA.

(b) Restriction enzyme digestion of cDNA and capture of 3' most NlaIII cDNA fragment.

(c) Addition of specific adaptors and excision of each cDNAs 10-bp tag.

(d) Formation of ditags, amplification and removal of adaptors.

(e) Formation of concateners and sequencing.

Drug Discovery Today

Figure 7.5
The Serial Analysis of Gene Expression. (Derived from Madden *et al.* [124].)

made single-stranded, then hybridized against a set of microbeads each separately containing a sequence complementary against a 32-mer tag. The microbeads are then fixed in place within a flow cell. Sequencing can take place directly on the microbeads through successive use of restriction enzymes exposing 4 base pairs, which in turn can be successively bound by florescently labeled probes.

Since the source templates can be generated from mRNA, this can effectively be used as a massively parallel method to find tags associated with each mRNA. Also, since presumably each copy of each mRNA is uniquely labeled with a different tag, each copy is sequenced separately, and thus this system is not only quantitative, but may be more sensitive at lower gene expression levels, like SAGE. At the time of writing, commercialization of this technique has just started.

7.1.6   Gel pad technology: Motorola

Using technology developed at the Argonne National Laboratory, Motorola Life Sciences (Northbrook, IL) has developed a novel "gel pad array" technology and has begun early production rounds of these arrays. Within each "gel pad," or gel matrix, several different kinds of reaction technologies can be embedded, including oligonucleotides. Although the oligonucleotides attached to the gel pads (themselves 100 μm in width) might not seem to have the same access to the free-floating complementary DNA strands in a biological sample as they might have on a glass slide, in practice they appear to have similar specificity and sensitivity of hybridization. The use of oligonucleotides is preferable to full length cDNA as reproducibility and cost are kept lower and the probes can be suitably normalized for GC content. As is the case for other oligonucleotide technologies, care must be taken in selecting an oligonucleotide sequence so that it does not cross-hybridize with other genes.

As the gel pads allow oligonucleotides lengths of greater than the 25μm used in Affymetrix microarrays, in theory they can provide higher specificity. Also, although the feature size is relatively large compared to oligonucleotide arrays made by Affymetrix, current gel pad technologies can be arrayed up to 10,000 probes on a single glass slide. As for other platforms, higher densities are promised.

7.2   Respecting the Older Generation

It has already happened that laboratories have invested thousands of dollars and hours into one generation of microarray technology, and then have had to move on to the next generation of microarrays. After several experiments, it becomes apparent to them that results across the generations of microarrays are not identical. The question then arises of how to salvage the information from the first generation of microarrays. The concern is, of course, that they will have to treat the data sets of different microarray generations as distinct and incomparable and therefore they will lose "statistical power" in their classification or clustering studies. This concern is magnified when the earlier studies involved unique and irreplaceable tissue samples which may no longer be available to be studied with the newer-generation microarrays. We illustrate this challenge here with some of our own experience with the Affymetrix platform. We hasten to add that the challenges we have had in combining results across generations of this platform have their correlates on other platforms.

7.2.1 The generation gap

The HuGeneFL and HG-U95A oligonucleotide microarrays from Affymetrix are two successive generations of the platform for human expression profiling. Because the latter microarray (one of a 5-chip set) is a superset of the former generation, one would expect that probe sets in one generation would have the same expression pattern as measured in the next generation of microarrays. However, as demonstrated in figure 7.6, there are a large number of probe sets that have very poor correlation when the same RNA sample is hybridized identically and in the same laboratory to the two generations of microarrays.

The most likely explanation for this disparity is that Affymetrix designed the HG-U95A array to use a set of probes or oligonucleotides for each probe set that may only partially overlap the set of probes used in the HuGeneFL arrays. Presumably this reflects improved oligonucleotide selections based on greater knowledge of the genome and accumulated empirical results of the specificity and sensitivity of particular probes. However, because Affymetrix has continued to maintain secrecy around the sequence of each oligonucleotide in a probe set, we cannot assess this directly. Affymetrix does release a table mapping the probes (within a probe set) in one generation to the next. This allows a comparison of how well probe sets correlate as a function of the number of shared probes (see figure 7.7). At a threshold of three or more probes in common per probe set, the correlation is substantially improved. This may explain in large part why the overall correlation between the two generations of microarrays is so poor.

7.2.2 Separating the wheat from the chaff

We know then that the consistency of expression measurements for the overall set of probe sets across a generation of microarrays is poor. There also is a subset of probe

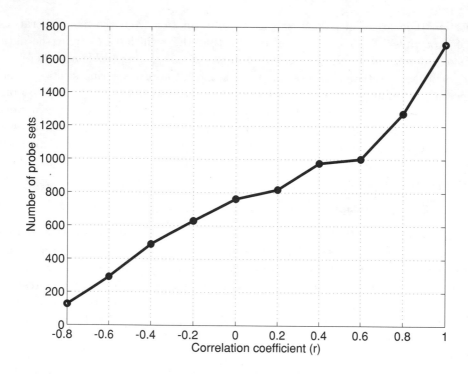

Figure 7.6
Poor measurement reproducibility across microarray generations. Two generations of Affymetrix microarrays were used to the gene expression of seven samples. Each of the 8075 common probe sets thus had seven measurements from both the older HuGeneFL and newer HG-U95A microarrays. The line represents the distribution of correlation coefficients of each probe set. Though the exact correlation coefficient may not be useful, it is notable that over 25% of the probe sets showed measurements that were zero or negatively correlated across generations. (Derived from Nimgaonkar *et al.* [136].)

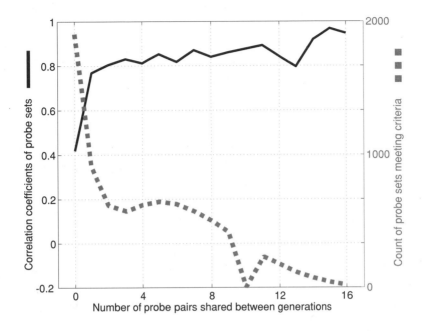

Figure 7.7
Measurement reproducibility of probe sets across microarray generations, by the number of
probe pairs in common. A total of 8075 probe sets are deemed in common by Affymetrix
between the older HuGeneFL and the newer HG-U95A microarrays. However, each probe set
has a varying number of common probe pairs, from none to all 16. The bar graph represents the
distribution of the 8075 probe sets, and the line represents the correlation coefficients of each set
of probe sets across generations. For example, one finds a correlation coefficient of .77 when
plotting all 6412 probe sets with one probe pair in common across microarray generations.
(Derived from Nimgaonkar *et al.* [136].)

sets for which consistency is high. To maximally salvage the data of the HuGeneFL arrays, then, the goal should be to determine which probe sets are members of that high-consistency subset. One reasonable heuristic is to pick a threshold correlation and and then pick only those probe sets that have at least n probes in common such that the correlation for n as obtained from the histogram of figure 7.7 is above the correlation threshold. Just what this threshold should be will be determined by a decision analytic procedure of the kind illustrated on page 2.9.

However, the above procedure will only provide a small subset of the genes represented by the various probe sets on each of the microarrays. There may be a substantially larger set of genes that are consistently reported across the generations of microarrays that might be "salvageable." If a particular laboratory is in possession of duplicate hybridizations across these generations, then the technique described in section 3.2.2 can be adapted to provide an estimate of which probe sets are the most consistent.

That is, if the identical RNA extract has been hybridized to two generations of microarrays, then a simple program can be written to determine for any measure of gene expression (*e.g.*, the Cy3/Cy5 ratio, Average Difference, Log Average Ratio, Present vs. Absent Calls, *etc.*) the threshold (*e.g.*, minimal Cy3/Cy5 ratio) above which the expression levels reported have an acceptable level of consistency across the microarray generations. These thresholds will not only vary with the particular microarray platform employed but also with the laboratory performing the hybridizations and the tissue types employed. Consequently, this procedure must be done for each laboratory and the results may not be safely generalizable across laboratories.

7.2.3 A persistent problem

This problem of the comparability of expression measurements across generations of microarrays is likely to be with us for several years. Microarray technology is in such flux that a definitive platform that only suffers small incremental changes is not in the immediate future. Given that laboratories will be studying the same biological system for several years, and therefore over several generations of microarrays, the most prudent and conservative advice would be to:

- Decouple the hybridization schedule from tissue sampling and banking. Preferably all the hybridizations should be done within a few months, if not days. The effect of the different time of hybridization is often visible in subsequent analysis. For example, careful review of the clustering of the acute myelogenous leukemia (AML) and acute lymphocytic leukemia (ALL) data set of Golub *et al.* [78] reveals that the test and training sets automatically cluster separately. That is, the differences between the gene profiles in the test and training sets are of the same order as the differences between AML and ALL. As reported in [78] the only difference between the two sets was the date on which the hybridization was done. In addition to environmental and procedural variation, it is likely that each batch of microarrays was manufactured on a different date.

- Purchase or construct sufficient microarrays to last your entire study. Even if a newer and better generation of microarrays becomes available, the quality of the study is best served by staying with a single generation of microarrays. Otherwise, the investigator will have to wander into the uncertain territory of estimating consistency across generations of microarray technology for each gene interrogated by the microarray, as outlined above. Unfortunately, some studies necessarily are so long as to make the logistics of this tactic unrealistic.

- Maintain a reference RNA pool. This can be used to determine the consistency of measurements across microarray generations when all else fails. The selection of the reference pool should be informed by the need to maintain a large amount of reference RNA over years with minimal degradation. In addition, the pool should contain a distribution of transcript abundance that is appropriate for the experiments envisaged. Too much or too little of a transcript will cause skewed noise or variation profiles due to different degrees of cross-hybridization against different biological samples.

7.3   Selecting Software

Any high-throughput functional genomics investigation will require that there be an armamentarium of analytic tools on hand. Even a large laboratory that has several bioinformaticians creating customized tools will still have to integrate these tools with the best-of-breed available commercially or through academic licenses. More often, there is insufficient time or personnel to develop any software locally, so external software tool kits must be obtained. The decision of how to go about acquiring these is inevitably weighed down by considerations of the local environment, local resources, budgetary constraints, and the need for customizability of these tool kits. It is well outside the scope of this book to address this process. Nonetheless, we provide here an abbreviated framework with which to consider acquisitions of bioinformatics software for use in functional genomics.

In order to provide the analytic services required, there are two broad categories of function that must be served: a collaborative data-sharing environment, and a set of robust analytic tools. The former will allow a research group to share common materials such as microarray data sets, annotations of these data sets, publications, and workflow tools such that the progress of any particular study can be monitored, and the latter will provide for the extraction of clinical or biological knowledge from genomic data sets.

Primary concerns in the first category, collaborative software platforms, will be

1. the scalability of the software to satisfy large numbers of local and remote users;

2. the security of the site to protect both the intellectual property of the data as well as the privacy of any patients whose data may be stored in the database.

A highly nontrivial aspect of the data-sharing or collaborative software tool kit is to maintain a data model for the microarrays that is capable of encompassing all the measurements that are likely to be obtained through the available microarray platforms (see section 5.4). Similarly, the collaborative software should have a sufficiently detailed and flexible phenotypic data model to allow structured annotations using controlled vocabularies to support the phenotypic-genotypic correlations that are an essential component of a successful functional genomics pipeline.

The second category of bioinformatics software includes the analytic tools that provide the implementation of the techniques that we have described in the data-mining chapter (chapter 4) and in the measurement techniques chapter (chapter 3) such as fold difference, self-organizing maps, dendrograms, support vector machines, relevance networks, k-means clustering, matrix incision trees, gene shaving, and others. They also provide means of visualizing analytic results and their relationships and linking these with the published literature, annotations, and links to relevant biomedical databases.

In the late 1990s software packages fell cleanly into one of those two aforementioned categories. Subsequently, in order to capture additional market share, the analytic tool vendors have added collaborative data-sharing capabilities and likewise the collaborative data sharing software has become increasingly equipped with a suite of analytic tools commonly used in the investigations reported in the literature. At present, however, few of the solutions are comprehensive. That is, the suite of analytic tools provided by the collaborative software vendors are limited or limiting in their analytic capabilities, and the collaborative, data-sharing, and warehousing facilities of the analytic tools are embryonic and of unproven scalability. Some of the best software packages are free; they have been developed often with federal funding by the best and the brightest and most public-minded of bioinformaticians. As a result, an investigator may not always get what he or she pays for in terms of value in quality software. Nonetheless, with respect to support for installation, debugging, and technical advice, these will be very limited with any of these free packages. Also, publicly available software packages typically require some knowledge of operating systems in order to be able to install them correctly on the computational hardware platforms. This is particularly true of Unix- and Linux-based computational platforms which are typically the most stable and reliable of platforms but also require the most expertise in technical support and management.

Some of the more notable free analysis solutions are listed and coarsely taxonomized in table 7.1. Note that many of these are available for free only to academic users, and charges may apply for commercial users. Also, refer to table 5.2 for free data models and databases.

Recent reviews of commercially available software tools in this domain include [29, 16, 35].

7.4   Investing in the Future of the Genomic Enterprise

If microarray technology is going to be commoditized such that it can be used routinely as a measurement system, even in relatively unskilled hands, then what will distinguish that particular genomic enterprise with aspirations after excellence?

The most pressing need and most obvious quality that will distinguish the best functional genomics investigations from the rest will be the nature of the investigators participating. As to which investigators these might be, the genomic pipeline outlined in section 1.5 points to the kinds of expertise that will be most in demand. There will need to be top-flight basic biologists, pathologists, clinicians with access to populations of interest, all capable of and motivated toward posing the high-yield clinical questions that are likely to drive many of these investigations. Based on the experience of several of the more successful laboratories, it seems that at least for the near future, the largest and most pressing need for expertise is in the area of bioinformatics. A typical successful laboratory will have approximately two thirds of its investigators drawn from the ranks of bioinformaticians. This proportion is likely to be maintained for at least the next 5 to 10 years, until the analytic tools are much more robust and the analytic approach is systematized and standardized. At that juncture, it may be that bioinformatic analyses will be packageable in the same way as common biostatistical analyses for clinical trials are packaged in pro-

Tool name	Description	URL http://
AMADA	Dendrograms, principal components analysis	web.hku.hk/~xxia/software/AMADA.htm
ANOVA	Matlab software programs for microarray data	www.jax.org/research/churchill/software/anova/
ArrayViewer	Differential gene expression	www.tigr.org/softlab
BRB array tools	Excel software, provides scatterplots, dendrograms, and class prediction	linus.nci.nih.gov/BRB-ArrayTools.html
CAGED	Clustering by expression dynamics	kebab.tch.harvard.edu/caged
Cleaver	K-means clustering, principal components analysis, and classification	classify.stanford.edu
Cluster and TreeView	Dendrograms, self-organizing maps, k-means clustering, principal components analysis	rana.lbl.gov/EisenSoftware.htm
Cyber-T	Differential gene expression	genomics.biochem.uci.edu/genex/cybert
dChip	Differential gene expression	www.dchip.org
Equalizer	Microarray normalization	organogenesis.ucsd.edu/TheEqualizer.htm
Expression Profiler	Dendrograms, with many dissimilarity measures	ep.ebi.ac.uk
GeneCluster	Self-organizing maps	www-genome.wi.mit.edu/MPR/GeneCluster/GeneCluster.html
J-Express	Dendrograms, self-organizing maps, principal components analysis, k-means clustering	www.ii.uib.no/~bjarted/jexpress/main.html
MAExplorer	Differential gene expression, scatterplots, k-means clustering, dendrograms	www-lecb.ncifcrf.gov/MAExplorer
Multiple Experiment Viewer	Many normalization, clustering, dissimilarity measures and graphical options	www.tigr.org/softlab
RelNet	Relevance networks	book.chip.org
SAM	Excel software, compares expression to clinical parameters	www-stat.stanford.edu/~tibs/SAM/index.html
ScanAlyze	Microarray spot detection	rana.lbl.gov/EisenSoftware.htm
SpotFinder	Microarray spot detection	www.tigr.org/softlab
XCluster	Dendrograms, self-organizing maps, k-means clustering	genome-www.stanford.edu/~sherlock/cluster.html

Table 7.1
List of freely available analysis software for microarrays

grams such as SAS (Statistical Analysis System) or SPSS (Statistical Package for the Social Sciences). Of course, the availability of such packages allows neophytes to perform uninformed, flawed, but superficially good-looking analyses, a problem that most biostatisticians are all too aware.

For the foreseeable future then, the most significant investments in expertise will probably have to be made in the area of bioinformatics. As individuals with training in biology and computer science and and medicine are rare and hard to come by, most laboratories are currently hiring individuals with strong computational skills, such as computational physicists, mathematicians, and statisticians, and providing them with the intellectual tools and background to become bioinformaticians. This process is not without its problems and uneven results.

A more lasting and productive process is to develop training programs that will produce the necessary interdisciplinary education in biology, computer science, statistics and probability theory, to mention just a few desirable knowledge bases that those individuals should be equipped with. The National Institutes of Health have recognized this need and are beginning to fund several training programs. Locally, the Division of Health, Sciences, and Technology (HST) at Harvard University and the Massachusetts Institute of Technology has developed a Bioinformatics and Integrative Genomics (BIG) training program. BIG is designed to take students with strong, quantitative undergraduate backgrounds such as chemical engineering, mechanical engineering, physics, mathematics, or computer science and then provide them with a curriculum that involves clinical exposure at Harvard Medical School during clinical rotations, joint classes with Harvard Medical School students, as well as basic biology courses. This and advanced topics in probability theory, computer science, and other genomics-related courses will provide these students with the basis to complete their doctoral thesis at HST in genomic science and subsequently to assume leadership roles in genomic research.

Even now, when many of the genomic measurement techniques have in fact not yet been commoditized, it is apparent that the most expensive and labor-intensive part of the genomic enterprise will be the accurate and large-scale phenotyping of human populations and model organisms. Specifically, it takes a lot more time and personnel to accurately document all the characteristics of a human pedigree with all the attendant clinical findings and measurements of each member of the pedigree, than to simply genotype or obtain expression measurements on each member of the pedigree. Consequently, any substantive forward-looking genomic effort will involve a carefully thought-out plan for the acquisition of detailed phenotypic information for all obtained samples. This activity has previously been most associated with genetic epidemiologists who have carefully obtained prospective perspectively gathered phenotypic data in well-designed studies such as the Nurses' Health Study, the Physicians' Health Study, and the Framingham Heart Study.

With motivations similar to those of epidemiologists, prospective annotation and phenotyping are indeed what are required. Those who expect to be able to mine the existing clinical records of a patient's care for the necessary data are going to be disappointed. Even if they can manage the issues of patient consent and data release, the quality of the clinical record and its emphasis is such that most of the data needed for effective and reliable phenotyping of the clinical population are unavailable. Consequently, any serious attempt to achieve high-throughput phenotyping for a substantive study will require that a laboratory or institution invest in the necessary infrastructure for such activities. These activities are labor-intensive and therefore not inexpensive. Specifically, it requires the use of trained personnel, such as genetic counselors, physician assistants, or highly trained research assistants, who are aware of the complexity of human phenotyping. As human subjects have relatively long lives as compared to microbial or murine models, these activities have to be supported adequately over decades. High-throughput human phenotyping also requires that the institutional laboratory put in place the consent and data release procedures that will enable the research to be done subsequently, and to ensure that these procedures are consistent with the aforementioned rules and regulations.

The importance of prospective phenotyping activity cannot be overemphasized. Without it, we will never be able to bridge the gap between the observed and measured phenotypic and genomic data and the clinical import of these data. This is a far less sexy activity than other aspects of the functional genomic enterprise, but it ultimately will be what distinguishes those efforts that are the most successful and clinically relevant. Who best performs such prospective high-throughput phenotyping will determine whether the locus of human-related genomics will be under the administrative ægis of hospitals, medical schools, the industrial sector, or governmental organizations.

called autosomes. Chromosome 1 is the largest and chromosome 22 the smallest. Each chromosome has two "arms" designated p and q.

Complementary DNA (cDNA) DNA that is synthesized from a messenger RNA template; the single-stranded form is often used as a probe in physical mapping or for detecting RNA. Since cDNA is constructed from messenger RNA (after introns have been spliced out), it does not contain introns.

Cosmid Artificially constructed cloning vector containing the *cos* gene of lambda phage (a virus). Cosmids can be packaged in lambda phage particles for infection into *E. coli*; this permits cloning of larger DNA fragments (up to 45 kb) than can be introduced into bacterial hosts in plasmid vectors.

Deoxyribonucleic Acid (DNA) The chemical that forms the basis of the genetic material in virtually all living organisms. Structurally, DNA is composed of two strands that intertwine to form a springlike structure called the double helix. Attached to each backbone are chemical structures called bases (or nucleotides), which protrude away from the backbone toward the center of the helix, and which come in four types—adenine, cytosine, guanine, and thymine (designated A, C, G, and T). In DNA, cytosine only forms optimal hydrogen bonding with guanine, and adenine only with thymine. These interactions across the many nucleotides in each strand hold the two strands together.

Electrophoresis The use of electrical fields to separate charged biomolecules such as DNA, RNA, and proteins. DNA and RNA carry a net negative charge because of the numerous phosphate groups in their structure. In the process of gel electrophoresis, these biomolecules are put into wells of a solid matrix typically made of an inert substance such as agarose. When this gel is placed into a bath and an electrical charge applied across the gel, the biomolecules migrate and separate according to size in proportion to the amount of charge they carry. The biomolecules can be stained for viewing and isolated and purified from the gels for further analysis. Electrophoresis can be used to isolate pure biomolecules from a mixture or to analyze biomolecules (such as for DNA sequencing).

Enzyme A protein that catalyzes chemical reactions in a living cell. Enzymes are protein molecules whose function it is to speed the making and breaking of chemical bonds required for essential physiochemical reactions.

Exon The protein-coding DNA sequences of a gene (see Intron).

Glossary

Constructed in part from the glossary in Primer on Molecular Genetics, U. S. Dept. of Energy, Human Genome Project, 1992.

Alternative Splicing Product Variation in the way pre-mRNA is spliced together. Introns are typically spliced from pre-mRNA and the remaining exons are pieced together to form a contiguous transcript of mRNA (see Exon). However, the same set of introns may not always be spliced out, and the resultant mRNA may have a different combination of introns and exons. These alternative versions of the mRNA result in different downstream proteins being formed (alternative splicing products). Alternative splicing products are only one reason why the number of gene products is much larger than the number of genes. It is also a reason why an oligonucleotide designed to hybridize with a subsegment of transcription products of a particular gene may fail to measure the transcription of that gene.

Base Pair The chemical structure that forms the units of DNA and RNA and that encode genetic information. The bases that make up the base pairs are adenine (A), guanine (G), thymine (T), cytosine (C), and uracil (U) (see DNA).

Binding Site Regions on macromolecules which are sequence and structure specific and complementary that enable any two or more macromolecules to mechanically bind and interact.

Bioinformatics The collection, organization, and analysis of large amounts of biological data, using networks of computers and databases. Historically, bioinformatics concerned itself with the analysis of the sequences of genes and their products (proteins), but the field has since expanded to the management, processing, analysis, and visualization of large quantities of data from genomics, proteomics, d screening, and medicinal chemistry. Bioinformatics also includes the integr and "mining" (detailed searching) of the ever-expanding databases of inforr from these disciplines.

Chromosome One of the physically separate segments that together genome, or total genetic material, of a cell. Chromosomes are long stran material, or DNA, that have been packaged and compressed by wra proteins. The number and size of chromosomes varies from specie humans, there are 23 pairs of chromosomes (a pair has one chrom parent). One pair are called the sex chromosomes because they determine sex. The chromosome carrying the male determinin Y and the corresponding female one is the X chromosome. T

Expressed Sequence Tags (EST) A set of single-pass sequenced cDNAs from an mRNA population derived from a specified cell population (*e.g.*, a tumor or a particular muscle type). ESTs provide one estimate of which DNA sequences are actually transcribed and therefore potentially functional as genes.

Functional genomics The use of genetic technology to determine the function of newly discovered genes by determining their role in one or more model organisms. Functional genomics uses as its starting point the isolated gene whose function is to be determined, and then selects a model organism in which a homolog of that gene exists. This model organism can be as simple as a yeast cell or as complex as a nematode worm, fruit fly, or even a mouse.

Gene The basic unit of heredity; the sequence of DNA that encodes all the information to make a protein. A gene may be "activated" or "switched on" to make protein—this activation is referred to as gene expression—by these proteins which control when, where, and how much protein is expressed from the gene. In the human genome, there are an estimated 30,000 genes (although recent studies suggest a larger number).

Gene product The biochemical material, either RNA or protein, resulting from expression of a gene. The amount of gene product is used to measure how active a gene is; abnormal amounts can be correlated with disease-causing alleles. As gene products include all the alternative splicing products, there are estimated to be at least 100,000 distinct such products.

Gene therapy The technology that uses genetic material for therapeutic purposes. This genetic material can be in the form of a gene, representative of a gene or cDNA, RNA, or even a small fragment of a gene. The introduced genetic material can be therapeutic in several ways: It can make a protein that is defective or missing in the patient's cells (as would be the case in a genetic disorder), or that will correct or modify a particular cellular function, or that elicits an immune response.

High-throughput screening The use of miniaturized, robotics-based technology to screen large compound libraries against an isolated target protein, cell, or tissue in order to identify binders that may be potential new drugs. In conjunction with genomics and combinatorial chemistry (the production of large numbers of medicinally relevant compounds), high-throughput screening has revolutionized the capacity of pharmaceutical and biotechnology companies to identify potential new drugs. Typically, high-throughput screening has relied on 96-well plates as the standard, although higher-density formats (384, 1536) are possible. Recently, advances in miniaturization and microfluidics have allowed screening of up to 100,000 compounds against a target on a single chip daily.

Hybridization The interaction of complementary nucleic acid strands. Since DNA is a double-stranded structure held together by complementary interactions (in which C always binds to G, and A to T), complementary strands favorably reanneal or "hybridize" to each other when separated.

Intron Noncoding portion of the gene that is spliced out from the nascent RNA transcript in the process of making an mRNA transcript. Frequently includes regulator elements (*i.e.*, binding sites) in addition to those of the promoter.

Microfluidics Microfluidics is the miniaturization of fluid-based biochemistry to very small volumes such that a large series of chemical reactions can be performed in parallel in a microarray format. The essential issue with developing microfluidics-based devices is the ability to maintain adequate access to the solutions needed in each element of the array. Several companies have manufactured "labs-on-a-chip" that allow the high-throughput chemistries needed in screening, synthesis, and probing of biological molecules.

Mutation Any alteration to DNA that can potentially result in a change in the function of one or more genes. Mutations can be a change in a single base of DNA (point mutation) or a loss of base pairs (deletion) affecting a single gene, or a movement of chromosomal regions (translocation) affecting many genes. Some changes in DNA occur naturally and lead to no harmful effects; these changes in a population are called polymorphisms.

Northern Blot RNA from a sample is spatially separated and distributed by mass on a gel. Radioactively labelled DNA or RNA strands with sequence complementary to the RNA segments from the sample are used to locate the position of those RNA segments.

Oligonucleotide A short molecule consisting of several linked nucleotides (typically between 10 and 60) chained together and attached by covalent bonds.

Open Reading Frame Regions in a nucleotide sequence that are bounded by start and stop codons and are therefore possible gene coding regions.

Pharmacogenomics Pharmacogenomics is the study of the pharmacological response to a drug by a population based on the genetic variation of that population.

Glossary

Constructed in part from the glossary in Primer on Molecular Genetics, U. S. Dept. of Energy, Human Genome Project, 1992.

Alternative Splicing Product Variation in the way pre-mRNA is spliced together. Introns are typically spliced from pre-mRNA and the remaining exons are pieced together to form a contiguous transcript of mRNA (see Exon). However, the same set of introns may not always be spliced out, and the resultant mRNA may have a different combination of introns and exons. These alternative versions of the mRNA result in different downstream proteins being formed (alternative splicing products). Alternative splicing products are only one reason why the number of gene products is much larger than the number of genes. It is also a reason why an oligonucleotide designed to hybridize with a subsegment of transcription products of a particular gene may fail to measure the transcription of that gene.

Base Pair The chemical structure that forms the units of DNA and RNA and that encode genetic information. The bases that make up the base pairs are adenine (A), guanine (G), thymine (T), cytosine (C), and uracil (U) (see DNA).

Binding Site Regions on macromolecules which are sequence and structure specific and complementary that enable any two or more macromolecules to mechanically bind and interact.

Bioinformatics The collection, organization, and analysis of large amounts of biological data, using networks of computers and databases. Historically, bioinformatics concerned itself with the analysis of the sequences of genes and their products (proteins), but the field has since expanded to the management, processing, analysis, and visualization of large quantities of data from genomics, proteomics, drug screening, and medicinal chemistry. Bioinformatics also includes the integration and "mining" (detailed searching) of the ever-expanding databases of information from these disciplines.

Chromosome One of the physically separate segments that together forms the genome, or total genetic material, of a cell. Chromosomes are long strands of genetic material, or DNA, that have been packaged and compressed by wrapping around proteins. The number and size of chromosomes varies from species to species. In humans, there are 23 pairs of chromosomes (a pair has one chromosome from each parent). One pair are called the sex chromosomes because they contain genes that determine sex.The chromosome carrying the male determining genes is designated Y and the corresponding female one is the X chromosome.The remaining pairs are called autosomes. Chromosome 1 is the largest and chromosome 22 the smallest. Each chromosome has two "arms" designated p and q.

Complementary DNA (cDNA) DNA that is synthesized from a messenger RNA template; the single-stranded form is often used as a probe in physical mapping or for detecting RNA. Since cDNA is constructed from messenger RNA (after introns have been spliced out), it does not contain introns.

Cosmid Artificially constructed cloning vector containing the *cos* gene of lambda phage (a virus). Cosmids can be packaged in lambda phage particles for infection into *E. coli*; this permits cloning of larger DNA fragments (up to 45 kb) than can be introduced into bacterial hosts in plasmid vectors.

Deoxyribonucleic Acid (DNA) The chemical that forms the basis of the genetic material in virtually all living organisms. Structurally, DNA is composed of two strands that intertwine to form a springlike structure called the double helix. Attached to each backbone are chemical structures called bases (or nucleotides), which protrude away from the backbone toward the center of the helix, and which come in four types—adenine, cytosine, guanine, and thymine (designated A, C, G, and T). In DNA, cytosine only forms optimal hydrogen bonding with guanine, and adenine only with thymine. These interactions across the many nucleotides in each strand hold the two strands together.

Electrophoresis The use of electrical fields to separate charged biomolecules such as DNA, RNA, and proteins. DNA and RNA carry a net negative charge because of the numerous phosphate groups in their structure. In the process of gel electrophoresis, these biomolecules are put into wells of a solid matrix typically made of an inert substance such as agarose. When this gel is placed into a bath and an electrical charge applied across the gel, the biomolecules migrate and separate according to size in proportion to the amount of charge they carry. The biomolecules can be stained for viewing and isolated and purified from the gels for further analysis. Electrophoresis can be used to isolate pure biomolecules from a mixture or to analyze biomolecules (such as for DNA sequencing).

Enzyme A protein that catalyzes chemical reactions in a living cell. Enzymes are protein molecules whose function it is to speed the making and breaking of chemical bonds required for essential physiochemical reactions.

Exon The protein-coding DNA sequences of a gene (see Intron).

Expressed Sequence Tags (EST) A set of single-pass sequenced cDNAs from an mRNA population derived from a specified cell population (*e.g.*, a tumor or a particular muscle type). ESTs provide one estimate of which DNA sequences are actually transcribed and therefore potentially functional as genes.

Functional genomics The use of genetic technology to determine the function of newly discovered genes by determining their role in one or more model organisms. Functional genomics uses as its starting point the isolated gene whose function is to be determined, and then selects a model organism in which a homolog of that gene exists. This model organism can be as simple as a yeast cell or as complex as a nematode worm, fruit fly, or even a mouse.

Gene The basic unit of heredity; the sequence of DNA that encodes all the information to make a protein. A gene may be "activated" or "switched on" to make protein—this activation is referred to as gene expression—by these proteins which control when, where, and how much protein is expressed from the gene. In the human genome, there are an estimated 30,000 genes (although recent studies suggest a larger number).

Gene product The biochemical material, either RNA or protein, resulting from expression of a gene. The amount of gene product is used to measure how active a gene is; abnormal amounts can be correlated with disease-causing alleles. As gene products include all the alternative splicing products, there are estimated to be at least 100,000 distinct such products.

Gene therapy The technology that uses genetic material for therapeutic purposes. This genetic material can be in the form of a gene, representative of a gene or cDNA, RNA, or even a small fragment of a gene. The introduced genetic material can be therapeutic in several ways: It can make a protein that is defective or missing in the patient's cells (as would be the case in a genetic disorder), or that will correct or modify a particular cellular function, or that elicits an immune response.

High-throughput screening The use of miniaturized, robotics-based technology to screen large compound libraries against an isolated target protein, cell, or tissue in order to identify binders that may be potential new drugs. In conjunction with genomics and combinatorial chemistry (the production of large numbers of medicinally relevant compounds), high-throughput screening has revolutionized the capacity of pharmaceutical and biotechnology companies to identify potential new drugs. Typically, high-throughput screening has relied on 96-well plates as the standard, although higher-density formats (384, 1536) are possible. Recently, advances in miniaturization and microfluidics have allowed screening of up to 100,000 compounds against a target on a single chip daily.

Hybridization The interaction of complementary nucleic acid strands. Since DNA is a double-stranded structure held together by complementary interactions (in which C always binds to G, and A to T), complementary strands favorably reanneal or "hybridize" to each other when separated.

Intron Noncoding portion of the gene that is spliced out from the nascent RNA transcript in the process of making an mRNA transcript. Frequently includes regulator elements (*i.e.*, binding sites) in addition to those of the promoter.

Microfluidics Microfluidics is the miniaturization of fluid-based biochemistry to very small volumes such that a large series of chemical reactions can be performed in parallel in a microarray format. The essential issue with developing microfluidics-based devices is the ability to maintain adequate access to the solutions needed in each element of the array. Several companies have manufactured "labs-on-a-chip" that allow the high-throughput chemistries needed in screening, synthesis, and probing of biological molecules.

Mutation Any alteration to DNA that can potentially result in a change in the function of one or more genes. Mutations can be a change in a single base of DNA (point mutation) or a loss of base pairs (deletion) affecting a single gene, or a movement of chromosomal regions (translocation) affecting many genes. Some changes in DNA occur naturally and lead to no harmful effects; these changes in a population are called polymorphisms.

Northern Blot RNA from a sample is spatially separated and distributed by mass on a gel. Radioactively labelled DNA or RNA strands with sequence complementary to the RNA segments from the sample are used to locate the position of those RNA segments.

Oligonucleotide A short molecule consisting of several linked nucleotides (typically between 10 and 60) chained together and attached by covalent bonds.

Open Reading Frame Regions in a nucleotide sequence that are bounded by start and stop codons and are therefore possible gene coding regions.

Pharmacogenomics Pharmacogenomics is the study of the pharmacological response to a drug by a population based on the genetic variation of that population.

It has long been known that different individuals in a population respond to the same drug differently, and that these variations are due to variations in the molecular receptors being affected by the drug, or to differences in metabolic enzymes that clear the drug. Pharmacogenomics is the science of studying these variations at the molecular level. Applications of pharmacogenomics include reducing side effects, customizing drugs, improvement of clinical trials, and the rescue of some drugs that have been banned due to severe side effects in a small percentage of the eligible population.

Polymorphism Difference in nucleotide sequence among individuals. Single nucleotide polymorphisms (SNP) are thought to be the principal component of inherited genetic variation. There is approximately one SNP every 300 to 1000 nucleotide of the human genome.

Probe Any biochemical agent that is labeled or tagged in some way so that it can be used to identify or isolate a gene, RNA, or protein. Typically refers to the immobilized specified nucleic acid in a detection system [8].

Promoter Regions of DNA typically upstream or before a gene which contain transcription factor binding sites (see Transcription Factor), but also include elements for specific initiation of transcription. That is, there are stretches of DNA to which transcription factors may bind and thereby modulate transcription but which are not promoters, because they do not include the elements of transcriptional initiation.

Protein family A set of proteins that share a common evolutionary origin reflected by their relatedness in function, which is usually reflected by similarities in sequence, or in primary, secondary, or tertiary structure. A set of proteins with related structure and function.

Proteomics The study of the entire protein complement or "protein universe" of the cell. Mirroring genomics, proteomics aims to determine the entire suite of expressed proteins in a cell. This includes determining the number, level, and turnover of all expressed proteins, their sequence and any post-translational modifications to the sequence, and protein-protein and protein-other molecule interactions within the cell, across the cell membrane, and among (secreted) proteins.

Polymerase chain reaction (PCR) A technique used to amplify or generate large amounts of replica DNA or a segment of any DNA whose "flanking" sequences are known. Oligonucleotide primers that bind these flanking sequences are used by an enzyme to copy the sequence in between the primers. Cycles of heat to break apart the DNA strands, cooling to allow the primers to bind, and heating again to allow the enzyme to copy the intervening sequence lead to doubling of the DNA present at each cycle.

Serial Analysis of Gene Expression (SAGE) The use of short 10-14 bp sequence tags linked together to identify sequences in a sample. The count of the instances of each sequence in a specific sample is a number that bears (in theory) a direct relationship to the number of transcripts of a particular genes (see section 7.1.4).

Single nucleotide polymorphism (SNP) Variations in single base pairs scattered throughout the human genome that serve as measures of genetic diversity in humans. About 1 million SNPs are estimated to be present in the human genome, and SNPs are useful markers for gene mapping and disease studies.

Southern Blot DNA from a sample is cut with restriction enzymes and the position of the fragments (*e.g.* on a gel) is determined by the fragment's molecular weight. Complementary strands of radioactively labelled DNA are used to identify the position of the DNA fragments on the gel.

Transcription Factor A molecute, typically a protein, which binds to promoter (see Promoter) or other DNA binding sites with some regulatory role in transcription. The binding (or unbinding) of a transcription factor from a promoter eventually leads to a change of transcription activity in the gene controlled by that promoter.

Target The free nucleic acid that is being interrogated [8].

References

[1] T. Akutsu, S. Miyano, and S. Kuhara. Algorithms for identifying Boolean networks and related biological networks based on matrix multiplication and fingerprint function. *Journal of computational biology*, 7(3-4):331–43, 2000.

[2] T. Akutsu, S. Miyano, and S. Kuhara. Inferring qualitative relations in genetic networks and metabolic pathways. *Bioinformatics*, 16(8):727–34, 2000.

[3] A. A. Alizadeh, M. B. Eisen, R. E. Davis, C. Ma, I. S. Lossos, A. Rosenwald, J. C. Boldrick, H. Sabet, T. Tran, X. Yu, J. I. Powell, L. Yang, G. E. Marti, T. Moore, J. Hudson, Jr., L. Lu, D. B. Lewis, R. Tibshirani, G. Sherlock, W. C. Chan, T. C. Greiner, D. D. Weisenburger, J. O. Armitage, R. Warnke, R. Levy, W. Wilson, M. R. Grever, J. C. Byrd, D. Botstein, P. O. Brown, and L. M. Staudt. Distinct types of diffuse large B-cell lymphoma identified by gene expression profiling. *Nature*, 403(6769):503–11, 2000.

[4] U. Alon, N. Barkai, D. A. Notterman, K. Gish, S. Ybarra, D. Mack, and A. J. Levine. Broad patterns of gene expression revealed by clustering analysis of tumor and normal colon tissues probed by oligonucleotide arrays. *Proceedings of the National Academy of Sciences of the United States of America*, 96(12):6745–50, 1999.

[5] O. Alter, P. O. Brown, and D. Botstein. Singular value decomposition for genome-wide expression data processing and modeling. *Proceedings of the National Academy of Sciences of the United States of America*, 97(18):10101–6, 2000.

[6] R. B. Altman and S. Raychaudhuri. Whole-genome expression analysis: challenges beyond clustering. *Current opinion in structural biology*, 11(3):340–7, 2001.

[7] Anonymous. Jargon file version 4.0.0, 1996.

[8] Anonymous. Array data go public. *Nature genetics*, 22(3):211–2, 1999.

[9] Anonymous. GEM microarray reproducibility study. Technical Report PN 99-0169, Incyte Pharmaceuticals, Inc., September 1999.

[10] Anonymous. *GeneChip analysis suite user guide*, volume version 3.3. Affymetrix, Inc., 1999.

[11] S. M. Arfin, A. D. Long, E. T. Ito, L. Tolleri, M. M. Riehle, E. S. Paegle, and G. W. Hatfield. Global gene expression profiling in Escherichia coli K12. The effects of integration host factor. *The Journal of biological chemistry*, 275(38):29672–84, 2000.

[12] A. Arkin, P. Shen, and J. Ross. A test case of correlation metric construction of a reaction pathway from measurements. *Science*, 277:1275–9, 1997.

[13] M. Ashburner, C. A. Ball, J. A. Blake, D. Botstein, H. Butler, J. M. Cherry, A. P. Davis, K. Dolinski, S. S. Dwight, J. T. Eppig, M. A. Harris, D. P. Hill, L. Issel-Tarver, A. Kasarskis, S. Lewis, J. C. Matese, J. E. Richardson, M. Ringwald, G. M. Rubin, and G. Sherlock. Gene Ontology: tool for the unification of biology. *Nature genetics*, 25(1):25–9, 2000.

[14] Roger Bacon. *Opus majus*. Russell and Russell, New York, 1962.

[15] J. Banfield and A. Raftery. Model-based Gaussian and non-Gaussian clustering. *Biometrics*, 49:803–821, 1993.

[16] D. E. Bassett, Jr., M. B. Eisen, and M. S. Boguski. Gene expression informatics–it's all in your mine. *Nature genetics*, 21(1 Suppl):51–5, 1999.

[17] L. R. Baugh, A. A. Hill, E. L. Brown, and C. P. Hunter. Quantitative analysis of mRNA amplification by in vitro transcription. *Nucleic acids research*, 29(5):E29, 2001.

[18] C. F. Belanger, C. H. Hennekens, B. Rosner, and F. E. Speizer. The nurses' health study. *The American journal of nursing*, 78(6):1039–40, 1978.

[19] A. Ben-Dor, L. Bruhn, N. Friedman, I. Nachman, M. Schummer, and Z. Yakhini. Tissue classification with gene expression profiles. *Journal of computational biology*, 7(3-4):559–83, 2000.

[20] A. Ben-Dor, R. Shamir, and Z. Yakhini. Clustering gene expression patterns. *Journal of computational biology*, 6(3-4):281–97, 1999.

[21] Amir Ben-Dor, Nir Friedman, and Zohar Yakhini. Tissue classification with gene expression profiles. In *RECOMB*, pages 31–38, Tokyo, Japan, 1999. ACM.

[22] D. A. Benson, I. Karsch-Mizrachi, D. J. Lipman, J. Ostell, B. A. Rapp, and D. L. Wheeler. GenBank. *Nucleic acids research*, 28(1):15–8, 2000.

[23] B. Berger and T. Leighton. Protein folding in the hydrophobic-hydrophilic (HP) model is NP-complete. *Journal of computational biology*, 5(1):27–40, 1998.

[24] F. Bertucci, K. Bernard, B. Loriod, Y. C. Chang, S. Granjeaud, D. Birnbaum, C. Nguyen, K. Peck, and B. R. Jordan. Sensitivity issues in DNA array-based expression measurements and performance of nylon microarrays for small samples. *Human molecular genetics*, 8(9):1715–22, 1999.

[25] V. E. Bichsel, L. A. Liotta, and E. F. Petricoin, 3rd. Cancer proteomics: from biomarker discovery to signal pathway profiling. *Cancer journal*, 7(1):69–78, 2001.

[26] M. Bittner, P. Meltzer, Y. Chen, Y. Jiang, E. Seftor, M. Hendrix, M. Radmacher, R. Simon, Z. Yakhini, A. Ben-Dor, N. Sampas, E. Dougherty, E. Wang, F. Marincola, C. Gooden, J. Lueders, A. Glatfelter, P. Pollock, J. Carpten, E. Gillanders, D. Leja, K. Dietrich, C. Beaudry, M. Berens, D. Alberts, and V. Sondak. Molecular classification of cutaneous malignant melanoma by gene expression profiling. *Nature*, 406(6795):536–40, 2000.

[27] M. Bittner, P. Meltzer, and J. Trent. Data analysis and integration: of steps and arrows. *Nature genetics*, 22(3):213–5, 1999.

[28] M. S. Boguski and G. D. Schuler. ESTablishing a human transcript map. *Nature genetics*, 10(4):369–71, 1995.

[29] D. D. Bowtell. Options available–from start to finish–for obtaining expression data by microarray. *Nature genetics*, 21(1 Suppl):25–32, 1999.

[30] A. Brazma and J. Vilo. Gene expression data analysis. *FEBS letters*, 480(1):17–24, 2000.

[31] S. Brenner, M. Johnson, J. Bridgham, G. Golda, D. H. Lloyd, D. Johnson, S. Luo, S. McCurdy, M. Foy, M. Ewan, R. Roth, D. George, S. Eletr, G. Albrecht, E. Vermaas, S. R. Williams, K. Moon, T. Burcham, M. Pallas, R. B. DuBridge, J. Kirchner, K. Fearon, J. Mao, and K. Corcoran. Gene expression analysis by massively parallel signature sequencing (MPSS) on microbead arrays. *Nature biotechnology*, 18(6):630–4, 2000.

[32] B. M. Broccolo and B. W. Petersen. Final HIPAA privacy rules: "How do we get started?". *Journal of health care finance*, 27(4):7–23, 2001.

[33] Michael P. S. Brown, William Noble Grundy, David Lin, Nello Cristianini, Charles Walsh Sugnet, Terrence S. Furey, Manuel Ares, Jr., and David Haussler. Knowledge-based analysis of microarray gene expression data by using support vector machines. *Proceedings of the National Academy of Sciences of the United States of America*, 97(1):262–267, 2000.

[34] P. O. Brown and D. Botstein. Exploring the new world of the genome with DNA microarrays. *Nature genetics*, 21(1 Suppl):33–7, 1999.

[35] Michael Brush. Making sense of microchip array data. *The Scientist*, 15(9):25, 2001.

[36] W. Buntine. Operations for learning with graphical models. *Journal of artificial intelligence research*, 2:159–225, 1994.

[37] A. J. Butte and I. S. Kohane. Unsupervised knowledge discovery in medical databases using relevance networks. In Nancy Lorenzi, editor, *Proceedings of the American Medical Informatics Association Fall Symposium*, pages 711–715, Washington, DC, 1999. Hanley & Belfus.

[38] A. J. Butte and I. S. Kohane. Mutual information relevance networks: functional genomic clustering using pairwise entropy measurements. In R. Altman, A. K. Dunker, L. Hunter, K. Lauderdale, and T. E. Klein, editors, *Pacific Symposium on Biocomputing*, volume 5, pages 418–29, Hawaii, 2000.

[39] A. J. Butte, P. Tamayo, D. Slonim, T. R. Golub, and I. S. Kohane. Discovering functional relationships between RNA expression and chemotherapeutic susceptibility using relevance networks. *Proceedings of the National Academy of Sciences of the United States of America*, 97(22):12182–6, 2000.

[40] Atul J. Butte, Jessica Ye, G. Niederfellner, K. Rett, H. U. Hädring, Morris F. White, and Issac S. Kohane. Determining significant fold differences in gene expression analysis. In R. Altman, A. K. Dunker, L. Hunter, K. Lauderdale, and T. E. Klein, editors, *Pacific Symposium on Biocomputing*, volume 6, pages 6–17, Hawaii, 2001.

[41] H. Caron, B. van Schaik, M. van der Mee, F. Baas, G. Riggins, P. van Sluis, M. C. Hermus, R. van Asperen, K. Boon, P. A. Voute, S. Heisterkamp, A. van Kampen, and R. Versteeg. The human transcriptome map: clustering of highly expressed genes in chromosomal domains. *Science*, 291(5507):1289–92, 2001.

[42] G. Casella and R. L. Berger. *Statistical inference*. Statistics and Probability. Brooks/Cole Publishing Company, Pacific Grove, CA, 1990.

[43] A. Chakravarti. Population genetics–making sense out of sequence. *Nature genetics*, 21(1 Suppl):56–60, 1999.

[44] T. Chen, H. L. He, and G. M. Church. Modeling gene expression with differential equations. In R. Altman, A. K. Dunker, L. Hunter, K. Lauderdale, and T. E. Klein, editors, *Pacific Symposium on Biocomputing*, volume 4, pages 29–40, Hawaii, 1999.

[45] Y. Chen, E. Dougherty, and M. Bittner. Ratio-based decisions and the quantitative analysis of cDNA microarray images. *Journal of biomedical optics*, 2(4):364–374, 1997.

[46] M. L. Chow, E. J. Moler, and I. S. Mian. Identifying marker genes in transcription profiling data using a mixture of feature relevance experts. *Physiological genomics*, 5(2):99–111, 2001.

[47] Paul D. Clayton, W. Earl Boebert, Gordon H. Defriese, Susan P. Dowell, Mary L. Fennell, Kathleen A. Frawley, John Glaser, Richard A. Kemmerer, Carl E. Landwehr, Thomas C. Rindfleisch, Sheila A. Ryan, Bruce J. Sams, Jr., Peter Szolovits, Robbie G. Trussell, and Elizabeth Ward. *For the record: protecting electronic health information.* National Academy Press, Washington, DC, 1997.

[48] E. F. Codd. A relational model of data for large shared data banks. *Communications of the ACM*, 13(6):377–387, 1970.

[49] E. Coiera. Clinical communication: a new informatics paradigm. In James Cimino, editor, *Proceedings of the American Medical Informatics Association Fall Symposium*, pages 17–21, Washington, DC, 1996. Hanley & Belfus.

[50] E. Coiera and V. Tombs. Communication behaviours in a hospital setting: an observational study. *BMJ*, 316(7132):673–6, 1998.

[51] M. H. Coletti and H. L. Bleich. Medical subject headings used to search the biomedical literature. *Journal of the American Medical Informatics Association*, 8(4):317–323, 2001.

[52] G. F. Cooper and E. Herskovitz. A bayesian method for the induction of probabilistic networks from data. *Machine learning*, 9:309–347, 1992.

[53] R. A. Cote and S. Robboy. Progress in medical information management: systematized nomenclature of medicine (SNOMED). *Journal of the American Medical Association*, 243:756–762, 1980.

[54] T. M. Cover and J. A. Thomas. *Elements of information theory.* Wiley-Interscience, New York, 1991.

[55] T. R. Dawber, G. F. Meadors, and F. E. J. Moore. The Framingham study: Epidemiological approaches to heart disease. *American journal of public health*, 41:279–286, 1951.

[56] J. DeRisi, L. Penland, P. O. Brown, M. L. Bittner, P. S. Meltzer, M. Ray, Y. Chen, Y. A. Su, and J. M. Trent. Use of a cDNA microarray to analyse gene expression patterns in human cancer. *Nature genetics*, 14(4):457–60, 1996.

[57] J. L. DeRisi, V. R. Iyer, and P. O. Brown. Exploring the metabolic and genetic control of gene expression on a genomic scale. *Science*, 278(5338):680–6, 1997.

[58] R. L. Devaney. *An introduction to chaotic dynamical systems.* Addison-Wesley, Redwood City, CA, 1989.

[59] T. G. Dietterich. Approximate statistical tests for comparing supervised classification learning algorithms. *Neural computation*, 10(7):1895–1923, 1998.

[60] Joaquin Dopazo, Edward Zanders, Ilaria Dragoni, Gillian Amphlett, and Francesco Falciani. Methods and approaches in the analysis of gene expression data. *Journal of immunological methods*, 250(1-2):93–112, 2001.

[61] Sandrine Dudoit. Statistical methods for identifying differentially expressed genes in replicated cDNA microarray experiments. Technical Report 578, University of California, Berkeley, August 2000.

[62] B. L. Ebert and H. F. Bunn. Regulation of transcription by hypoxia requires a multiprotein complex that includes hypoxia-inducible factor 1, an adjacent transcription factor, and p300/CREB binding protein. *Molecular and cellular biology*, 18(7):4089–96, 1998.

[63] M. B. Eisen, P. T. Spellman, P. O. Brown, and D. Botstein. Cluster analysis and display of genome-wide expression patterns. *Proceedings of the National Academy of Sciences of the United States of America*, 95(25):14863–8, 1998.

[64] D. Endy and R. Brent. Modelling cellular behaviour. *Nature*, 409 Suppl(6818):391–5, 2001.

[65] O. Ermolaeva, M. Rastogi, K. D. Pruitt, G. D. Schuler, M. L. Bittner, Y. Chen, R. Simon, P. Meltzer, J. M. Trent, and M. S. Boguski. Data management and analysis for gene expression arrays. *Nature genetics*, 20(1):19–23, 1998.

[66] R. M. Ewing, A. B. Kahla, O. Poirot, F. Lopez, S. Audic, and J. M. Claverie. Large-scale statistical analyses of rice ESTs reveal correlated patterns of gene expression. *Genome research*, 9(10):950–9, 1999.

[67] L. A. Farrer, L. A. Cupples, J. L. Haines, B. Hyman, W. A. Kukull, R. Mayeux, R. H. Myers, M. A. Pericak-Vance, N. Risch, and C. M. van Duijn. Effects of age, sex, and ethnicity on the association between apolipoprotein E genotype and Alzheimer disease. *Journal of the American Medical Association*, 278(16):1349–56, 1997.

[68] O. Fiehn, J. Kopka, P. Dormann, T. Altmann, R. N. Trethewey, and L. Willmitzer. Metabolite profiling for plant functional genomics. *Nature biotechnology*, 18(11):1157–61, 2000.

[69] D. Figeys and D. Pinto. Proteomics on a chip: promising developments. *Electrophoresis*, 22(2):208–16, 2001.

[70] S. P. Fodor, J. L. Read, M. C. Pirrung, L. Stryer, A. T. Lu, and D. Solas. Light-directed, spatially addressable parallel chemical synthesis. *Science*, 251(4995):767–73, 1991.

[71] N. Friedman, M. Linial, I. Nachman, and D. Pe'er. Using bayesian networks to analyze expression data. *Journal of computational biology*, 7(3-4):601–20, 2000.

[72] T. S. Furey, N. Cristianini, N. Duffy, D. W. Bednarski, M. Schummer, and D. Haussler. Support vector machine classification and validation of cancer tissue samples using microarray expression data. *Bioinformatics*, 16(10):906–14, 2000.

[73] T. Gaasterland and S. Bekiranov. Making the most of microarray data. *Nature genetics*, 24(3):204–6, 2000.

[74] F. Gamarra, G. Simic-Schleicher, R. M. Huber, A. Ulsenheimer, P. C. Scriba, U. Kuhnle, and M. Wehling. Impaired rapid mineralocorticoid action on free intracellular calcium in pseudohypoaldosteronism. *The Journal of clinical endocrinology and metabolism*, 82(3):831–4, 1997.

[75] G. K. Geiss, R. E. Bumgarner, M. C. An, M. B. Agy, A. B. van 't Wout, E. Hammersmark, V. S. Carter, D. Upchurch, J. I. Mullins, and M. G. Katze. Large-scale monitoring of host cell gene expression during HIV-1 infection using cDNA microarrays. *Virology*, 266(1):8–16, 2000.

[76] G. Getz, E. Levine, and E. Domany. Coupled two-way clustering analysis of gene microarray data. *Proceedings of the National Academy of Sciences of the United States of America*, 97(22):12079–84, 2000.

[77] W. Wayt Gibbs. Shrinking to enormity. *Scientific American*, 284(2):33–34, February 2001.

[78] T. R. Golub, D. K. Slonim, P. Tamayo, C. Huard, M. Gaasenbeek, J. P. Mesirov, H. Coller, M. L. Loh, J. R. Downing, M. A. Caligiuri, C. D. Bloomfield, and E. S. Lander. Molecular classification of cancer: class discovery and class prediction by gene expression monitoring. *Science*, 286(5439):531–7, 1999.

[79] S. Greenfield, S. Cretin, L. G. Worthman, and F. Dorey. The use of an ROC curve to express quality of care results. *Medical decision making*, 2(1):23–31, 1982.

[80] J. Guckenheimer and P. Holmes. *Nonlinear oscillations, dynamical systems, and bifurcations of vector fields*, volume 42 of *Applied Mathematical Sciences*. Springer-Verlag, New York, 1983.

[81] I. Guyon, J. Weston, S. Barnhill, and V. Vapnik. Gene selection for cancer classification using support vector machines. *Machine learning*, 46(1-3):389–422, 2002.

[82] Ira J. Haimowitz, Ramesh S. Patil, and Peter Szolovits. Representing medical knowledge in a terminological language is difficult. In R. A. Greenes, editor, *Proceedings of the Twelfth Symposium on Computer Applications in Medical Care*, pages 101–105, Washington, DC, 1988. IEEE Computer Society Press.

[83] J. Hanke and J. G. Reich. Kohonen map as a visualization tool for the analysis of protein sequences: multiple alignments, domains and segments of secondary structures. *Computer applications in the biosciences*, 12(6):447–54, 1996.

[84] A. Harding and C. Stuart-Buttle. The development and role of the Read Codes. *Journal of American Health Information Management Association*, 69(5):34–8, 1998.

[85] A. J. Hartemink, D. K. Gifford, T. S. Jaakkola, and R. A. Young. Using graphical models and genomic expression data to statistically validate models of genetic regulatory networks. In R. Altman, A. K. Dunker, L. Hunter, K. Lauderdale, and T. E. Klein, editors, *Pacific Symposium on Biocomputing*, volume 6, pages 422–33, Hawaii, 2001.

[86] T. Hastie, R. Tibshirani, D. Botstein, and P. Brown. Supervised harvesting of expression trees. *Genome biology*, 2(1), 2001.

[87] T. Hastie, R. Tibshirani, M. B. Eisen, A. Alizadeh, R. Levy, L. Staudt, W. C. Chan, D. Botstein, and P. Brown. "gene shaving" as a method for identifying distinct sets of genes with similar expression patterns. *Genome biology*, 1(2), 2000.

[88] D. Heckerman, D. Geiger, and D. M. Chickering. Learning bayesian networks: The combination of knowledge and statistical data. Technical Report MSR-TR-94-09, Microsoft Research, March 1994.

[89] David Heckerman. Bayesian networks for data mining. *Data mining and knowledge discovery*, 1(1):79–119, 1997.

[90] I. Hedenfalk, D. Duggan, Y. Chen, M. Radmacher, M. Bittner, R. Simon, P. Meltzer, B. Gusterson, M. Esteller, O. P. Kallioniemi, B. Wilfond, A. Borg, and J. Trent. Gene-expression profiles in hereditary breast cancer. *The New England journal of medicine*, 344(8):539–48, 2001.

[91] A. Herbert and A. Rich. RNA processing and the evolution of eukaryotes. *Nature genetics*, 21(3):265–9, 1999.

[92] J. Herrero, A. Valencia, and J. Dopazo. A hierarchical unsupervised growing neural network for clustering gene expression patterns. *Bioinformatics*, 17(2):126–36, 2001.

[93] Ralf Herwig, Albert J. Poustka, Christine Muller, Christof Bull, Hans Lehrach, and John O'Brien. Large-scale clustering of cDNA-fingerprinting data. *Genome research*, 9:1093–1105, 1999.

[94] L. J. Heyer, S. Kruglyak, and S. Yooseph. Exploring expression data: identification and analysis of coexpressed genes. *Genome research*, 9(11):1106–15, 1999.

[95] S. G. Hilsenbeck, W. E. Friedrichs, R. Schiff, P. O'Connell, R. K. Hansen, C. K. Osborne, and S. A. Fuqua. Statistical analysis of array expression data as applied to the problem of tamoxifen resistance. *Journal of the National Cancer Institute*, 91(5):453–9, 1999.

[96] K. Hokamp and K. Wolfe. What's new in the library? What's new in GenBank? Let PubCrawler tell you. *Trends in genetics*, 15(11):471–2, 1999.

[97] F. C. Holstege, E. G. Jennings, J. J. Wyrick, T. I. Lee, C. J. Hengartner, M. R. Green, T. R. Golub, E. S. Lander, and R. A. Young. Dissecting the regulatory circuitry of a eukaryotic genome. *Cell*, 95(5):717–28, 1998.

[98] Yuh-Jyh Hu. An integrated approach for genome-wide gene expression analysis. *Computer methods and programs in biomedicine*, 65(3):163–174, 2001.

[99] S. M. Huff, R. A. Rocha, C. J. McDonald, G. J. De Moor, T. Fiers, W. D. Bidgood, Jr., A. W. Forrey, W. G. Francis, W. R. Tracy, D. Leavelle, F. Stalling, B. Griffin, P. Maloney, D. Leland, L. Charles, K. Hutchins, and J. Baenziger. Development of the Logical Observation Identifier Names and Codes (LOINC) vocabulary. *Journal of the American Medical Informatics Association*, 5(3):276–92, 1998.

[100] M. Ishii, S. Hashimoto, S. Tsutsumi, Y. Wada, K. Matsushima, T. Kodama, and H. Aburatani. Direct comparison of GeneChip and SAGE on the quantitative accuracy in transcript profiling analysis. *Genomics*, 68(2):136–43, 2000.

[101] A. K. Jain and R. C. Dubes. *Algorithms for clustering data*. Prentice Hall, Englewood Cliffs, NJ, 1988.

[102] K. K. Jain. Tech.Sight. Biochips for gene spotting. *Science*, 294(5542):621–3, 2001.

[103] M. Janowitz. A relational approach to ordinal clustering and classification. 2000. In preparation.

[104] R. A. Jungmann, D. Huang, and D. Tian. Regulation of LDH-A gene expression by transcriptional and posttranscriptional signal transduction mechanisms. *The Journal of experimental zoology*, 282(1-2):188–95, 1998.

[105] R. E. Kass and A. Raftery. Bayes factors. *Journal of the American Statistical Association*, 90:773–795, 1995.

[106] M. K. Kerr, M. Martin, and G. A. Churchill. Analysis of variance for gene expression microarray data. *Journal of computational biology*, 7(6):819–37, 2000.

[107] J. Khan, J. S. Wei, M. Ringner, L. H. Saal, M. Ladanyi, F. Westermann, F. Berthold, M. Schwab, C. R. Antonescu, C. Peterson, and P. S. Meltzer. Classification and diagnostic prediction of cancers using gene expression profiling and artificial neural networks. *Nature medicine*, 7(6):673–9, 2001.

[108] Ju Han Kim, Lucila Ohno-Machado, and Isaac S. Kohane. Unsupervised learning from complex data: the matrix incision tree algorithm. In R. Altman, A. K. Dunker, L. Hunter, K. Lauderdale, and T. E. Klein, editors, *Pacific Symposium on Biocomputing*, volume 6, pages 30–41, Hawaii, 2001.

[109] P. E. Kloeden and E. Platen. *Numerical solution of stochastic differential equations*, volume 23 of *Applications of Mathematics*. Springer-Verlag, New York, 1992.

[110] G. T. Klus, A. Song, A. Schick, M. Wahde, and Z. Szallasi. Mutual information analysis as a tool to assess the role of aneuploidy in the generation of cancer-associated differential gene expression patterns. In R. Altman, A. K. Dunker, L. Hunter, K. Lauderdale, and T. E. Klein, editors, *Pacific Symposium on Biocomputing*, volume 6, pages 42–51, Hawaii, 2001.

[111] Isaac S. Kohane, Hongmei Dong, and Peter Szolovits. Health information identification and de-identification toolkit. In Christopher Chute, editor, *Proceedings of the American Medical Informatics Association Fall Symposium*, pages 356–360, Philadelphia, PA, 1998. Hanley & Belfus.

[112] L. Kruglyak. Prospects for whole-genome linkage disequilibrium mapping of common disease genes. *Nature genetics*, 22(2):139–44, 1999.

[113] Winston P. Kuo, Tor-Kristian Jenssen, Atul J. Butte, L. Ohno-Machado, and Isaac S. Kohane. Analysis of matched mRNA measurements from two different microarray technologies. *Bioinformatics*, 18(3):405–412, 2002.

[114] R. H. Lathrop. The protein threading problem with sequence amino acid interaction preferences is NP-complete. *Protein engineering*, 7(9):1059–68, 1994.

[115] S. L. Lauritzen. *Graphical models*. Oxford University Press, New York, 1996.

[116] C. K. Lee, R. G. Klopp, R. Weindruch, and T. A. Prolla. Gene expression profile of aging and its retardation by caloric restriction. *Science*, 285(5432):1390–3, 1999.

[117] C. K. Lee, R. Weindruch, and T. A. Prolla. Gene-expression profile of the ageing brain in mice. *Nature genetics*, 25(3):294–297, 2000.

[118] G. Lennon, C. Auffray, M. Polymeropoulos, and M. B. Soares. The I.M.A.G.E. Consortium: an integrated molecular analysis of genomes and their expression. *Genomics*, 33(1):151–2, 1996.

[119] S. Liang, S. Fuhrman, and R. Somogyi. Reveal, a general reverse engineering algorithm for inference of genetic network architectures. In R. Altman, A. K. Dunker, L. Hunter, K. Lauderdale, and T. E. Klein, editors, *Pacific Symposium on Biocomputing*, volume 3, pages 18–29, Hawaii, 1998.

[120] D. A. B. Lindberg and B. L. Humphreys. The Unified Medical Language System (UMLS) and computer-based patient records. In M. J. Ball and M. F. Collen, editors, *Aspects of the computer-based patient record*, pages 165–175. Springer-Verlag, New York, 1992.

[121] R. J. Lipshutz, S. P. Fodor, T. R. Gingeras, and D. J. Lockhart. High density synthetic oligonucleotide arrays. *Nature genetics*, 21(1 Suppl):20–4, 1999.

[122] D. J. Lockhart, H. Dong, M. C. Byrne, M. T. Follettie, M. V. Gallo, M. S. Chee, M. Mittmann, C. Wang, M. Kobayashi, H. Horton, and E. L. Brown. Expression monitoring by hybridization to high-density oligonucleotide arrays. *Nature biotechnology*, 14(13):1675–80, 1996.

[123] L. Luo, R. C. Salunga, H. Guo, A. Bittner, K. C. Joy, J. E. Galindo, H. Xiao, K. E. Rogers, J. S. Wan, M. R. Jackson, and M. G. Erlander. Gene expression profiles of laser-captured adjacent neuronal subtypes. *Nature medicine*, 5(1):117–22, 1999.

[124] S. L. Madden, C. J. Wang, and G. Landes. Serial analysis of gene expression: from gene discovery to target identification. *Drug discovery today*, 5(9):415–425, 2000.

[125] K. D. Mandl, P. Szolovits, and I. S. Kohane. Public standards and patients' control: how to keep electronic medical records accessible but private. *BMJ*, 322(7281):283–7, 2001.

[126] P. M. Mannucci. Polymorphisms in the factor VII gene and the risk of myocardial infarction. *The New England journal of medicine*, 344(6):458–9, 2001.

[127] D. R. Masys, J. B. Welsh, J. Lynn Fink, M. Gribskov, I. Klacansky, and J. Corbeil. Use of keyword hierarchies to interpret gene expression patterns. *Bioinformatics*, 17(4):319–26, 2001.

[128] H. Matsuno, A. Doi, M. Nagasaki, and S. Miyano. Hybrid petri net representation of gene regulatory network. In R. Altman, A. K. Dunker, L. Hunter, K. Lauderdale, and T. E. Klein, editors, *Pacific Symposium on Biocomputing*, volume 5, pages 341–52, Hawaii, 2000.

[129] D. B. McCarn. MEDLINE: An introduction to on-line searching. *Journal of the American Society for Information Science*, 31(3):181–192, 1980.

[130] Clement J. McDonald, L. Blevins, W. M. Tierney, and D .K. Martin. The Regenstrief medical records. *MD Computing*, 5(5):34–47, 1988.

[131] R. McEntire, P. Karp, N. Abernethy, D. Benton, G. Helt, M. DeJongh, R. Kent, A. Kosky, S. Lewis, D. Hodnett, E. Neumann, F. Olken, D. Pathak, P. Tarczy-Hornoch, L. Toldo, and T. Topaloglou. An evaluation of ontology exchange languages for bioinformatics. In *Proceedings of the International Conference on Intelligent Systems for Molecular Biology*, volume 8, pages 239–50, San Diego, 2000.

[132] A. A. Mironov, J. W. Fickett, and M. S. Gelfand. Frequent alternative splicing of human genes. *Genome research*, 9(12):1288–93, 1999.

[133] E. J. Moler, D. C. Radisky, and I. S. Mian. Integrating naive bayes models and external knowledge to examine copper and iron homeostasis in S. cerevisiae. *Physiological genomics*, 4(2):127–135, 2000.

[134] Kevin Murphy and Saira Mian. Modelling gene expression data using dynamic bayesian networks. Technical report, Lawrence Berkeley National Laboratory, 1999.

[135] M. A. Newton, C. M. Kendziorski, C. S. Richmond, F. R. Blattner, and K. W. Tsui. On differential variability of expression ratios: improving statistical inference about gene expression changes from microarray data. *Journal of computational biology*, 8(1):37–52, 2001.

[136] A. Nimgaonkar, D. Sanoudou, A. J. Butte, J. N. Haslett, L. M. Kunkel, A. H. Beggs, and I. S. Kohane. Reproducibility of gene expression across generations of Affymetrix microarrays. 2002. Submitted.

[137] R. Patil, R. Fikes, P. Patel-Schneider, D. McKay, T. Finin, T. Gruber, and R. Neches. The DARPA knowledge sharing effort: Progress report. In *Principles of Knowledge Representation and Reasoning: Third International Conference*, Royal Sonesta Hotel, Cambridge, MA, 1992.

[138] M. E. Patti, X. J. Sun, J. C. Bruening, E. Araki, M. A. Lipes, M. F. White, and C. R. Kahn. 4PS/insulin receptor substrate (IRS)-2 is the alternative substrate of the insulin receptor in IRS-1-deficient mice. *The Journal of biological chemistry*, 270(42):24670–3, 1995.

[139] J. Pearl. *Probabilistic reasoning in intelligent systems: networks of plausible inference*. Morgan Kaufmann Publishers, San Mateo, CA, 1988.

[140] J. Pearl. Causal diagrams for empirical research. *Biometrika*, 82(4):669–710, 1995.

[141] A. Persidis. Proteomics. *Nature biotechnology*, 16(4):393–4, 1998.

[142] D. M. Pisanelli, A. Gangemi, and G. Steve. WWW-available conceptual integration of medical terminologies: the ONIONS experience. In Daniel Masys, editor, *Proceedings of the American Medical Informatics Association Fall Symposium*, pages 575–579, Nashville, TN, 1997. Hanley & Belfus.

[143] K. R. Popper. *The logic of scientific discovery*. Basic Books, New York, 1959.

[144] W. H. Press, S. A. Teukolsky, W. T. Vetterling, and B. P. Flannery. *Numerical recipes: the art of scientific computing*. Cambridge University Press, New York, 2nd edition, 1992.

[145] K. D. Pruitt and D. R. Maglott. RefSeq and LocusLink: NCBI gene-centered resources. *Nucleic acids research*, 29(1):137–40, 2001.

[146] J. R. Quinlan. *C4.5: programs for machine learning*. Morgan Kaufmann Publishers, San Mateo, CA, 1992.

[147] J. Quinn. An HL7 (Health Level Seven) overview. *Journal of American Health Information Management Association*, 70(7):32–4, 1999.

[148] M. Ramoni and P. Sebastiani. Bayesian methods. In M. Berthold and D. J. Hand, editors, *Intelligent Data Analysis. An Introduction*, pages 129–166. Springer, New York, NY, 1999.

[149] S. Raychaudhuri, J. M. Stuart, and R. B. Altman. Principal components analysis to summarize microarray experiments: application to sporulation time series. In R. Altman, A. K. Dunker, L. Hunter, K. Lauderdale, and T. E. Klein, editors, *Pacific Symposium on Biocomputing*, volume 5, pages 455–66, Hawaii, 2000.

[150] B. Y. Reis, A. Butte, and I. S. Kohane. Extracting knowledge from dynamics in gene expression. *Journal of biomedical informatics*, 34(1):15–27, 2001.

[151] K. Reue. mRNA quantitation techniques: considerations for experimental design and application. *The Journal of nutrition*, 128(11):2038–44, 1998.

[152] C. S. Richmond, J. D. Glasner, R. Mau, H. Jin, and F. R. Blattner. Genome-wide expression profiling in Escherichia coli K-12. *Nucleic acids research*, 27(19):3821–3835, 1999.

[153] David M. Rind, Isaac S. Kohane, Peter Szolovits, Charles Safran, Henry C. Chueh, and G. Octo Barnett. Maintaining the confidentiality of medical records shared over the internet and world wide web. *Annals of internal medicine*, 127(2):138–141, 1997.

[154] D. T. Ross, U. Scherf, M. B. Eisen, C. M. Perou, C. Rees, P. Spellman, V. Iyer, S. S. Jeffrey, M. Van de Rijn, M. Waltham, A. Pergamenschikov, J. C. Lee, D. Lashkari, D. Shalon, T.G. Myers, J. N. Weinstein, D. Botstein, and P. O. Brown. Systematic variation in gene expression patterns in human cancer cell lines. *Nature genetics*, 24(3):227–35, 2000.

[155] D. E. Rumelhart and J. L. McClelland. *Parallel distributed processing: explorations in the microstructure of cognition.* MIT Press, Cambridge, MA, 1986.

[156] E. E. Schadt, C. Li, C. Su, and W. H. Wong. Analyzing high-density oligonucleotide gene expression array data. *Journal of cellular biochemistry,* 80(2):192–202, 2000.

[157] M. Schena, D. Shalon, R. W. Davis, and P. O. Brown. Quantitative monitoring of gene expression patterns with a complementary DNA microarray. *Science,* 270(5235):467–70, 1995.

[158] P. L. Schuyler, W. T. Hole, M. S. Tuttle, and D. D. Sherertz. The UMLS Metathesaurus: representing different views of biomedical concepts. *Bulletin of the Medical Library Association,* 81(2):217–22, 1993.

[159] J. A. Segal, J. L. Barnett, and D. L. Crawford. Functional analyses of natural variation in Sp1 binding sites of a TATA-less promoter. *Journal of molecular evolution,* 49(6):736–49, 1999.

[160] C. E. Shannon. A mathematical theory of communication. *Bell Systems Technical Journal,* 27:623–58, 1948.

[161] A. Shenker, L. Laue, S. Kosugi, J. J. Merendino, Jr., T. Minegishi, and G. B. Cutler, Jr. A constitutively activating mutation of the luteinizing hormone receptor in familial male precocious puberty. *Nature,* 365(6447):652–4, 1993.

[162] G. Sherlock. Analysis of large-scale gene expression data. *Current opinion in immunology,* 12(2):201–5, 2000.

[163] A. N. Shiryaev. *Probability,* volume 95 of *Graduate Texts in Mathematics.* Springer-Verlag, Cambridge, MA, 2nd edition, 1996.

[164] R. Somogyi and C. A. Sniegoski. Modeling the complexity of genetic networks: understanding multigenic and pleiotropic regulation. *Complexity,* 1(6):45–63, 1996.

[165] A. Soukas, P. Cohen, N. D. Socci, and J. M. Friedman. Leptin-specific patterns of gene expression in white adipose tissue. *Genes & development,* 14(8):963–80, 2000.

[166] E. Southern, K. Mir, and M. Shchepinov. Molecular interactions on microarrays. *Nature genetics,* 21(1 Suppl):5–9, 1999.

[167] P. T. Spellman, G. Sherlock, M. Q. Zhang, V. R. Iyer, K. Anders, M. B. Eisen, P. O. Brown, D. Botstein, and B. Futcher. Comprehensive identification of cell cycle-regulated genes of the yeast Saccharomyces cerevisiae by microarray hybridization. *Molecular biology of the cell,* 9(12):3273–97, 1998.

[168] D. J. Spiegelhalter and S. L. Lauritzen. Sequential updating of conditional probabilities on directed graphical structures. *Networks,* 20:157–224, 1990.

[169] Latanya Sweeney. Replacing personally-identifying information in medical records, the SCRUB system. In James Cimino, editor, *Proceedings of the American Medical Informatics Association Fall Symposium,* pages 333–337, Washington, DC, 1996. Hanley & Belfus.

[170] Latanya Sweeney. Guaranteeing anonymity when sharing medical data, the Datafly system. In Daniel Masys, editor, *Proceedings of the American Medical Informatics Association Fall Symposium,* pages 51–55, Nashville, TN, 1997. Hanley & Belfus.

[171] Latanya Sweeney. Three computational systems for disclosing medical data in the year 1999. In B. Cesnik, A. T. McCray, and J. R. Scherrer, editors, *Medinfo*, volume 9 Pt 2, pages 1124–9, Seoul, Korea, 1998. IOS Press.

[172] Z. Szallasi and S. Liang. Modeling the normal and neoplastic cell cycle with "realistic Boolean genetic networks": their application for understanding carcinogenesis and assessing therapeutic strategies. In R. Altman, A. K. Dunker, L. Hunter, K. Lauderdale, and T. E. Klein, editors, *Pacific Symposium on Biocomputing*, volume 3, pages 66–76, Hawaii, 1998.

[173] P. Szolovits and S. G. Pauker. Categorical and probabilistic reasoning in medical diagnosis. *Artificial Intelligence*, 11:115–144, 1978.

[174] Peter Szolovits and Isaac Kohane. Against simple universal health identifiers. *Journal of the American Medical Informatics Association*, 1(4):316–319, 1994.

[175] P. Tamayo, D. Slonim, J. Mesirov, Q. Zhu, S. Kitareewan, E. Dmitrovsky, E. S. Lander, and T. R. Golub. Interpreting patterns of gene expression with self-organizing maps: methods and application to hematopoietic differentiation. *Proceedings of the National Academy of Sciences of the United States of America*, 96(6):2907–12, 1999.

[176] S. Tavazoie, J. D. Hughes, M. J. Campbell, R. J. Cho, and G. M. Church. Systematic determination of genetic network architecture. *Nature genetics*, 22(3):281–5, 1999.

[177] D. Thieffry and R. Thomas. Qualitative analysis of gene networks. In R. Altman, A. K. Dunker, L. Hunter, K. Lauderdale, and T. E. Klein, editors, *Pacific Symposium on Biocomputing*, volume 3, pages 77–88, Hawaii, 1998.

[178] Robert Tibshirani, Guenther Walther, and Trevor Hastie. Estimating the number of clusters in a dataset via the gap statistic. Technical report, Department of Statistics, Stanford University, March 2000.

[179] M. Tomita, K. Hashimoto, K. Takahashi, T. S. Shimizu, Y. Matsuzaki, F. Miyoshi, K. Saito, S. Tanida, K. Yugi, J. C. Venter, and C. A. Hutchison, III. E-CELL: software environment for whole-cell simulation. *Bioinformatics*, 15(1):72–84, 1999.

[180] P. Toronen, M. Kolehmainen, G. Wong, and E. Castren. Analysis of gene expression data using self-organizing maps. *FEBS letters*, 451(2):142–6, 1999.

[181] C. L. Tsien, T. A. Libermann, X. Gu, and I. S. Kohane. On reporting fold differences. In R. Altman, A. K. Dunker, L. Hunter, K. Lauderdale, and T. E. Klein, editors, *Pacific Symposium on Biocomputing*, volume 6, pages 496–507, Hawaii, 2001.

[182] E. P. van Someren, L. F. Wessels, and M. J. Reinders. Linear modeling of genetic networks from experimental data. In *Proceedings of the International Conference on Intelligent Systems for Molecular Biology*, volume 8, pages 355–66, 2000.

[183] V. E. Velculescu, L. Zhang, B. Vogelstein, and K. W. Kinzler. Serial analysis of gene expression. *Science*, 270(5235):484–7, 1995.

[184] V. E. Velculescu, L. Zhang, W. Zhou, J. Vogelstein, M. A. Basrai, D. E. Bassett, Jr., P. Hieter, B. Vogelstein, and K. W. Kinzler. Characterization of the yeast transcriptome. *Cell*, 88(2):243–51, 1997.

[185] M. Wahde and J. Hertz. Modeling genetic regulatory dynamics in neural development. *Journal of computational biology*, 8(4):429–42, 2001.

[186] D. C. Weaver, C. T. Workman, and G. D. Stormo. Modeling regulatory networks with weight matrices. In R. Altman, A. K. Dunker, L. Hunter, K. Lauderdale, and T. E. Klein, editors, *Pacific Symposium on Biocomputing*, volume 4, pages 112–23, Hawaii, 1999.

[187] M. C. Weinstein, H. V. Fineberg, A. S. Elstein, H. S. Frazier, D. Neuhauser, R. R. Neutra, and B. J. McNeil. *Clinical decision analysis*. W. B. Saunders, Philadelphia, 1980.

[188] Sholom M. Weiss and Nitin Indurkhya. *Predictive data mining: a practical guide*. Morgan Kaufmann Publishers, San Francisco, 1997.

[189] X. Wen, S. Fuhrman, G. S. Michaels, D. B. Carr, S. Smith, J. L. Barker, and R. Somogyi. Large-scale temporal gene expression mapping of central nervous system development. *Proceedings of the National Academy of Sciences of the United States of America*, 95(1):334–9, 1998.

[190] J. Whittaker. *Graphical models in applied multivariate statistics*. Wiley, New York, 1990.

[191] L. D. Wilsbacher and J. S. Takahashi. Circadian rhythms: molecular basis of the clock. *Current opinion in genetics & development*, 8(5):595–602, 1998.

[192] D. J. Withers, D. J. Burks, H. H. Towery, S. L. Altamuro, C. L. Flint, and M. F. White. Irs-2 coordinates Igf-1 receptor-mediated beta-cell development and peripheral insulin signalling. *Nature genetics*, 23(1):32–40, 1999.

[193] Peter J. Woolf and Yixin Wang. A fuzzy logic approach to analyzing gene expression data. *Physiological genomics*, 3(1):9–15, 2000.

[194] S. Wright. Correlation and causation. *Journal of agricultural research*, 20:557–585, 1921.

[195] S. Wright. The theory of path coefficients: a reply to Niles' criticism. *Genetics*, 8:239–255, 1923.

[196] S. Wright. The method of path coefficients. *Annals of mathematical statistics*, 5:161–215, 1934.

[197] A. Wuensche. Genomic regulation modeled as a network with basins of attraction. In R. Altman, A. K. Dunker, L. Hunter, K. Lauderdale, and T. E. Klein, editors, *Pacific Symposium on Biocomputing*, volume 3, pages 89–102, Hawaii, 1998.

[198] J. J. Wyrick, F. C. Holstege, E. G. Jennings, H. C. Causton, D. Shore, M. Grunstein, E. S. Lander, and R. A. Young. Chromosomal landscape of nucleosome-dependent gene expression and silencing in yeast. *Nature*, 402(6760):418–21, 1999.

[199] Yee Hwa Yang, Michael J. Buckley, Sandrine Dudoit, and Terence P. Speed. Comparison of methods for image analysis on cDNA microarray data. Technical Report 584, Univeristy of California, Berkeley, 2000.

[200] D. B. Young. A post-genomic perspective. *Nature medicine*, 7(1):11–3, 2001.

Index

Page numbers in *italics* are illustrations.